VIRAL REGULATORY STRUCTURES AND THEIR DEGENERACY

VIRAL REGULATORY STRUCTURES AND THEIR DEGENERACY

Editor

Gerald Myers

Los Alamos National Laboratory
Los Alamos, New Mexico USA

Proceedings Volume XXVIII

Santa Fe Institute
Studies in the Sciences of Complexity

The Advanced Book Program

CRC Press
Taylor & Francis Group
Boca Raton   London   New York

CRC Press is an imprint of the
Taylor & Francis Group, an **informa** business

Director of Publications, Santa Fe Institute: *Ronda K. Butler-Villa*
Production Manager, Santa Fe Institute: *Della L. Ulibarri*
Publication Assistant/Indexer, Santa Fe Institute: *Marylee Thomson*

First published 1998 by Addison Wesley Longman, Inc.

Published 2018 by CRC Press
Taylor & Francis Group
6000 Broken Sound Parkway NW, Suite 300
Boca Raton, FL 33487-2742

CRC Press is an imprint of the Taylor & Francis Group, an informa business

Visit the Taylor & Francis Web site at
http://www.taylorandfrancis.com

and the CRC Press Web site at
http://www.crcpress.com

ISBN 13: 978-0-201-32822-6 (pbk)
ISBN 13: 978-0-201-32821-9 (hbk)

This volume was typeset using T$_{\mathrm{E}}$Xtures on a PowerMac 7200/120 computer.

About the Santa Fe Institute

The *Santa Fe Institute* (SFI) is a private, independent, multidisciplinary research and education center, founded in 1984. Since its founding, SFI has devoted itself to creating a new kind of scientific research community, pursuing emerging science. Operating as a small, visiting institution, SFI seeks to catalyze new collaborative, multidisciplinary projects that break down the barriers between the traditional disciplines, to spread its ideas and methodologies to other individuals, and to encourage the practical applications of its results.

All titles from the *Santa Fe Institute Studies in the Sciences of Complexity* series will carry this imprint which is based on a Mimbres pottery design (circa A.D. 950–1150), drawn by Betsy Jones. The design was selected because the radiating feathers are evocative of the outreach of the Santa Fe Institute Program to many disciplines and institutions.

Contributors to This Volume

Androphy, Elliot J., Department of Dermatology, New England Medical Center, and Department of Molecular Biology and Microbiology, Tufts University, School of Medicine, Boston, MA 02111

Baker, Carl C., Basic Research Laboratory, Division of Basic Sciences, National Institutes of Health, 41 Library Dr. MSC 5055, Bethesda, MD 20892-5055

Chao, Yesu, Howard Hughes Medical Institute, Departments of Medicine, Microbiology and Immunology, University of California–San Francisco, San Francisco, CA 94143-0724

Cujec, Thomas P., Howard Hughes Medical Institute, Departments of Medicine, Microbiology and Immunology, University of California–San Francisco, San Francisco, CA 94143-0724

Cullen, Bryan R., Howard Hughes Medical Institute and Department of Genetics, Duke University Medical Center, Durham, NC 27710

Ellington, Andrew D., Department of Chemistry, Indiana University, Bloomington, IN 47405-4001

Felber, Barbara K., ABL-Basic Research Program, Bldg. 535, Rm. 110, NCI-FCRDC, P.O. Box B, Frederick, MD 21702-1201

Frankel, Alan D., Department of Biochemistry and Biophysics, University of California, San Francisco, CA 94143-0448

Hitchcock, Penelope J., Sexually Transmitted Diseases Branch National Institute of Allergy and Infectious Diseases National Institutes of Health Bethesda, MD 20892

Huynen, Martijn, Biocomputing, EMBL, Meyerhofstrasse 1, 6900 Heidelberg, GERMANY and Santa Fe Institute, 1399 Hyde Park Road, Santa Fe, NM 87501, USA

Jeang, Kuan-Teh, Molecular Virology Section, National Institute of Allergy and Infectious Diseases, National Institutes of Health, Bethesda, MD 20892-0460

Konings, Danielle, Department of Microbiology, Southern Illinois University at Carbondale, Carbondale, Il 62901, USA and Santa Fe Institute, 1399 Hyde Park Road, Santa Fe, NM 87501, USA

McBride, Alison A., Laboratory of Viral Diseases, National Institute of Allergy and Infectious Diseases, National Institutes of Health, Bethesda, MD 20892-0455

Meyer, Jon K., Howard Hughes Medical Institute, Departments of Medicine, Microbiology and Immunology, University of California–San Francisco, San Francisco, CA 94143-0724

Münger, Karl, Department of Pathology, Harvard Medical School, Boston, MA 02115-5701

Myers, Gerald, Theoretical Biology and Biophysics, Los Alamos National Laboratory, T-10, MS K710, Los Alamos, NM 87545

Neuveut, Christine, Molecular Virology Section, National Institute of Allergy and Infectious Diseases, National Institutes of Health, Bethesda, MD 20892–0460

Okamoto, Hiroshi, Howard Hughes Medical Institute, Departments of Medicine, Microbiology and Immunology, University of California–San Francisco, San Francisco, CA 94143–0724

Pavlakis, George N., Human Retrovirus Section, NCI-FCRDC, P.O. Box B, Frederick, MD 21702-1201

Peterlin, B. Matija, Howard Hughes Medical Institute, Departments of Medicine, Microbiology and Immunology, University of California–San Francisco, San Francisco, CA 94143–0724

Schuster, Peter, Institut für Theoretische Chemie, Universität Wien, Währingerstraße 17, A-1090 Wien, AUSTRIA and Santa Fe Institute, 1399 Hyde Park Road, Santa Fe, NM 87501

Somogyi, Roland, National Institutes of Health, Laboratory of Neurophysiology, NINDS, Building 36/2C02, Bethesda, MD 20892

Stadler, Peter F., Institut für Theoretische Chemie, Universität Wien, Währingerstraße 17, A-1090 Wien, AUSTRIA and Santa Fe Institute, 1399 Hyde Park Road, Santa Fe, NM 87501

Thierry, Françoise, Unité des Virus Oncogènes, U1644 CNRS, Institut Pasteur, 28 rue du Dr Roux, 75724, Paris, Cedex 15, FRANCE

Yaniv, Moshe, Unité des Virus Oncogènes, U1644 CNRS, Institut Pasteur, 28 rue du Dr Roux, 75724, Paris, Cedex 15, FRANCE

Wain-Hobson, Simon, URM, Pasteur Institute, 25 Rue du Dr. Roux, 75724 Paris Cedex 15, FRANCE

Santa Fe Institute
Studies in the Sciences of Complexity

Proceedings Volumes

Lectures Volumes

Vol.	Editor	Title
I	D. L. Stein	Lectures in the Sciences of Complexity, 1989
II	E. Jen	1989 Lectures in Complex Systems, 1990
III	L. Nadel & D. L. Stein	1990 Lectures in Complex Systems, 1991
IV	L. Nadel & D. L. Stein	1991 Lectures in Complex Systems, 1992
V	L. Nadel & D. L. Stein	1992 Lectures in Complex Systems, 1993
VI	L. Nadel & D. L. Stein	1993 Lectures in Complex Systems, 1995

Lecture Notes Volumes

Vol.	Author	Title
I	J. Hertz, A. Krogh, & R. Palmer	Introduction to the Theory of Neural Computation, 1991
II	G. Weisbuch	Complex Systems Dynamics, 1991
III	W. D. Stein & F. J. Varela	Thinking About Biology, 1993
IV	J. M. Epstein	Nonlinear Dynamics, Mathematical Biology, and Social Science, 1997
V	H. F. Nijhout, L. Nadel, & D. L. Stein	Pattern Formation in the Physical and Biological Sciences, 1997

Reference Volumes

Vol.	Author	Title
I	A. Wuensche & M. Lesser	The Global Dynamics of Cellular Automata: Attraction Fields of One-Dimensional Cellular Automata, 1992

Contents

Preface

The National Institute of Allergy and Infectious Diseases (NIAID) has a profound commitment to supporting research that will improve the health of people. In the last two decades, a plethora of genetic information has become available. Many viruses and some bacterial genomes have been completely sequenced, as well as an enormous number of human genes. The challenge facing the biomedical community is to maximize the impact of this information on science, medicine and health. In 1986, the NIAID established an HIV (human immunodeficiency virus) Sequence Database and more recently an HPV (human papillomavirus) Sequence Database. These databases have been extremely useful in organizing and making use of the wealth of information (i.e., scientific opportunity) inherent in genetic sequences.

The exploration of the molecular similarities and differences of HIV and HPV, or of sexually transmitted pathogens in general, is a natural and critically important extension of these efforts. From the nucleic acid sequences, the amino acid sequences of the gene products can be inferred, and from studies of primary, secondary, and tertiary structure, functional relationships can be inferred. As one considers the next level of complexity—host/pathogen relationships—a third dimension of understanding is possible. Indeed, structure-function relationships are inherent within and across biological systems. An understanding of those complex relationships is likely to have a profound impact on the development and delivery of biomedical tools to prevent and/or control infection and disease. Some examples include the early detection of (1) the biopathways pathognomic for particular manifestations of infection, (2) the propensity for disease progression, (3) the rate of disease progression, and (4) the development of specific therapeutic interventions.

This book arose out of a workshop on HIV and HPV Regulatory Structures sponsored by the National Institute of Allergy and Infectious Diseases and held at the Santa Fe Institute in Santa Fe, New Mexico, May 6–8, 1996. Approximately 40 virologists and theoreticians from several countries were invited to examine experimental findings and models regarding HIV and HPV regulatory mechanisms. It was unusual to bring together researchers from the two areas; HIV and HPV infect different cells and display vastly different evolutionary potentials, one being a rapidly changing RNA virus, the other a slowly changing DNA virus. Nevertheless, they both possess small, compact genomes, and there are similarities in their control of transcriptional and posttranscriptional regulation that virologists typically have little opportunity to compare. Whether or not these viruses are capable of communicating in the intracellular environment is important since, taking into account epidemiologic data that is predicted by the reproductive rate of sexually transmitted infections, it is likely that all HIV-infected persons are coinfected with HPV. The dissimilarities in HIV and HPV regulatory strategies are also illuminating. Also, it was desirable to bring experimentalists together with theoretical and computational biologists who are interested in genetic processes and accustomed, at the Santa Fe Institute, to cross-disciplinary conversation.

We are indebted to Andi Sutherland, Ronda Butler-Villa, Della Ulibarri, and Marylee Thomson of the Santa Fe Institute for their help with the workshop and this book.

Penelope J. Hitchcock
D.V.M., M.S. Branch Chief
Sexually Transmitted Diseases Branch
National Institute of Allergy and Infectious Diseases
National Institutes of Health

Gerald Myers
Theoretical Biology and Biophysics
Los Alamos National Laboratory

November 3, 1997

Gerald Myers[†] and George N. Pavlakis[‡]

[†]Theoretical Biology and Biophysics, Los Alamos National Laboratory, T-10, MS K710, Los Alamos New Mexico 87545; E-mail: glm@lanl.gov

[†]The Santa Fe Institute, 1399 Hyde Park Road, Santa Fe New Mexico, 87501

[‡]Human Retrovirus Section, NCI-FCRDC, P.O. Box B, Frederick, MD 21702-1201; E-mail: PAVLAKIS@FCRFV2.NCIFCRF.GOV

Introduction

If viruses were merely the foreign toxic agents that their name suggests (the word "virus" is derived from a Latin word that connotes poison and sliminess) then most aspects of their natural history could be described by relatively straightforward dose-dependent curves and we would relegate the study of their properties to the disciplines of toxicology and molecular pharmacology. Viruses can, of course, destroy cells and tissues: many examples can be cited in evidence of how viruses become intricately antagonistic to the cell's defenses.[28,63] Thus, as a recent popular book describes them, they are "invaders—[which] cause more sickness than anything else on earth."[69] But, interestingly enough, the same book also describes them as "co-travelers in life," which suggests a fundamental ambivalence about their nature. Indeed, viruses can be viewed as adaptations of self, thus as Lewis Thomas suggested some years ago "viruses, instead of being single-minded agents of disease and death, now begin to look more like mobile genes."[89] Thomas's insight is astonishing given that it was reached without our contemporary knowledge that a relatively large fraction of the human genome is viral in origin.[44] Furthermore, we now know about the inclusion of human genes in viral genomic molecules (oncogenes being the most

Viral Regulatory Structures and Their Degeneracies, edited by Gerald Myers.
SFI Studies in the Sciences of Complexity, Proc. Vol. XXVIII, Addison-Wesley, 1998

1

studied cases).[3,16] For us, the classic question of whether viruses are living or not has been supplanted by the question of whether viruses are self or (like bacteria) clearly other.

One controversial but suggestive evolutionary theory proposes that pathogenicity from viruses is a microecological aberration rather than the rule.[95] This is not to say that diseases are few or insignificant, or that viral pathogenicity does not under certain circumstances increase through evolutionary selection. But the evolutionary perspective accords well with the notion that viral regulatory properties are copies or mimics of the essential control elements of the cell being inhabited; hence viruses serve as prisms for our observation and understanding of normal cellular regulatory processes. It is the factors that appear to promote commensalism, rather than pathogenesis, that will be examined for the most part in this book.

Viruses illustrate how slight perturbations can have major consequences. This now familiar pattern of complexity is shown clearly in the case of the small viruses HIV (human immunodeficiency virus) and HPV (human papillomavirus), which are the principal subjects of this book. These viruses can cause disease, HIVs more so than HPVs probably because they have had much less time to adapt to human hosts.[56,58] Yet many forms of immunodeficiency viruses and papillomaviruses are relatively benign, slowly insinuating themselves into cells with little damage. This suggests a tendency toward symbiosis.

The view that viruses are essentially self involves some complex problems. It would seem, on the basis of our line of thinking, that a virus would strive to adapt to its host by carefully limiting its collection of regulatory elements. It may then be asked why some viruses, for example vaccinia, are large, nearly as large and complex as some bacteria.[55] However, most viruses are small, though it is not at all clear that parsimony accounts for this fact. It might also seem as though the association of a virus with slow disease could signify a certain symbiotic strategy, but at least in the case of HIV this is deceptive: the viral replication dynamics and killing capacity turn out to be surprisingly high.[11,66] HIV and papillomaviruses could be less "single-minded" than they are, and they could be a lot larger. Be that as it may, we shall see that their handful of structural elements are more than sufficient to produce recycling effects characteristic of complexity in natural networks.[29]

We will return to the broader questions of viral regulation later in this introduction. In the sections that follow, the immunodeficiency viruses and papillomaviruses will be discussed and the chapters of this book will be introduced. The primary focus will be on the nature, origins, and degeneracy (or redundancy) of regulatory elements and on the strategies that enable viruses to adapt to cells.

1. HIV

Retroviruses are a large family of RNA viruses possessing reverse transcriptase and integrase enzymes, which enable them to transcribe their RNA genome into DNA that can be integrated into the host chromosome. In this host chromosomal phase, the viral particle is called a provirus and is said to be latent[10,13]; the latency refers to replication but not to transcription. The human immunodeficiency viruses were discovered in 1983, when the search for the causative agent of AIDS became urgent and, from some points of view, embarrassingly competitive. The initial thought was that HIV was closely related to recently discovered leukemia-causing retroviruses, but it was quickly shown through electron microscopy that they were members of another genus, the lentiviruses. Lenti-retroviruses (lenti means slow in Latin) have many molecular, biological, and clinical characteristics in common that are summarized elsewhere.[13,43,57] For the purposes of this introduction, they can be said to be complex retroviruses that cause a variety of syndromes with long incubation period and chronic active clinical picture but are not themselves typically oncogenic. They are associated with immunodeficiency, arthritic, and autoimmune conditions as well as central nervous system lesions. The essential lentiviral characteristics are that they can infect nondividing cells[60,64] and that they persist for long periods despite robust immune responses.

So-called complex retroviruses are larger retroviruses (with about 10 genes) that regulate their own expression, which simpler retroviruses with a minimal retroviral gene set cannot do[57] (see Felber). Lentiviral relatives of HIV are immunodeficiency viruses separately found in monkeys, cats, and cattle, Visna and Jembrana viruses that infect sheep, caprine arthritis encephalitis virus, and equine infectious anemia virus. Since the discovery of HIV, researchers have thought for various reasons that a murine lentivirus should exist, yet this viral type has never been recovered in nature. Foamy retroviruses and oncoretroviruses are also complex in the sense of owning regulatory genes that can function to regulate themselves. As we shall see, the regulatory capabilities of these three complex retroviral genera extend to one another, in spite of highly evolved dissimilarities.

The exact mechanism of HIV pathogenesis is not known, and many possibilities continue to be put forward.[9,39,43,99] By one school of thought, AIDS is regarded to be an autoimmune complication of what should otherwise be a benign state; hence, superantigenicity may be a factor.[36] Infected cells explain the pathogenesis for the most part, but with some remaining controversies and problems. The level of virus expression correlates with disease development in humans but not with HIV in chimpanzees nor with lentiviruses in other animal species. Simian immunodeficiency viruses (SIV), very close relatives to HIV which have virtually the same genetic composition and regulatory capabilities, are found in large numbers of apparently healthy green monkeys in the wild.[38] Furthermore, it is extremely difficult to cause AIDS-like disease in captive chimpanzees,[38,75] suggesting that the pathogenesis of HIV in humans is the result of a recent cross-species transmission event from animals highly adapted to the virus. In view of the widespread lack of pathogenicity—as distinct from viremia—in wild animals, it is probably safest to separate in one's mind the regulatory effects of immunodeficiency viruses from the pathogenic consequences of HIV infection.

It is well established that HIV has a tropism for actively dividing hematopoietic cells possessing a certain major receptor called CD4, and that the rate of decline in host CD4+ cells is directly tied to viral load.[11] Macrophages and T-lymphocytes are prime examples of cells susceptible to CD4-mediated HIV infection; among other cells infected by HIV are follicular dendritic cells, brain astrocytes, hepatocytes, and bowel epithelium. The CD4 molecule is not a requirement for viral entry into some cells (e.g., neural and epithelial cells), however, and when it is, secondary receptors undoubtedly play a role in uptake.[20,21] While HIV can infect terminally differentiated macrophages, the most productive infections appear with T-lymphocytes and macrophage precursors, pointing to the need for involvement of intracellular factors.[43,64] To begin to understand these capabilities and interactions, it will be helpful to introduce the proteins and elements of HIV, placing the focus of attention upon the regulatory features.

Like all retroviruses, HIV has Gag, Pol, and envelope proteins. The Gag polyprotein serves many structural functions associated with incorporation of the viral genome into a budding viral particle. Although it has self-associating properties, it requires extensive modification by host enzymes. In human, but not simian, immunodeficiency viruses, assistance is required from chaperone molecules such as cyclophilin A, which can be recovered from purified virions. Curiously, SIVs do not appear to use cyclophilin A, suggesting that other chaperones are involved in their particle formation. Gag is cleaved by a virally encoded protease in order to generate the matrix, capsid, and nucleocapsid components of the viral core. In addition, some small peptides with unknown service functions are produced through cleavage. A relatively large Gag-Pol fusion protein is formed through a low-frequency frameshifting event in the translation of the adjacent *gag* and *pol* genes. Control of this event requires a RNA structure known as a pseudoknot (see Huynen and Konings, hereafter H&K), to ensure that neither too much nor too little enzyme is produced and that the timing of the production is optimal.

One of the Pol enzymes is the protease needed for Gag and envelope cleavage. Other *pol*-encoded enzymes are the reverse transcriptase/ribonuclease H and the integrase, which are essential for synthesis of the provirus, transport to the cellular nucleus, and integration into the host chromosome. The Gag matrix protein as well as a viral factor Vpr (discussed below) may assist the nuclear import of the double-stranded provirus, although it is now believed that the Gag-Pol precursor, sometimes called assemblin, may itself confer this property to the particle after cell entry. The much-studied HIV envelope is a highly variable protein that serves mainly as the coat and binding factor for the infectious viral particle, hence the envelope recognizes the CD4+ receptor, as well as other essential coreceptors and alternative cell receptors, and mediates cell entry. The transmembrane portion of the envelope has fusigenic properties, which give rise *in vitro* to cellular clusters known as syncytia. The significance of syncytium formation *in vivo* is unclear. It is the extensive modification of the envelope protein through cellular enzymes (glycosylation) and its remarkable genetic variability that have been primary impediments to an HIV vaccine.

Together, the Gag, Pol, and envelope proteins define the retroviral life cycle: in simple outline, free virus attaches to a cell and penetrates by virtue of the envelope and appropriate cell receptors; the viral capsid is degraded, the RNA genome is reverse transcribed by viral enzyme that accompanies it in the virion, and the provirus is transported to the nucleus by virtue of Gag and Vpr proteins, also carried by the virion; upon integration the provirus can be transcribed; viral messages (transcripts) are transported back to the cytoplasm where proteins are manufactured; finally, new particles are packaged and bud from the cell.[64] Unintegrated viral DNA is found in cells, but productive transcription is thought to require the integrated proviral stage. We infer from human and viral sequences that on very rare occasions the provirus will deposit genetic material into the host or take genes away when it leaves.[13,16] Although, heterologous recombination has not been documented for HIV, as the AIDS epidemic unfolds, the possibility for genetic exchange and acquisition of new genes theoretically increases.[92]

The simplest retrovirus can achieve the above-outlined steps, probably with the aid of cellular factors that facilitate nuclear transport of the viral preintegration complex, transcription, and nuclear export of transcripts. HIV enhances its life cycle, first, by virtue of a protein called Tat (for transactivation), which has become an essential regulatory element for the virus. Once proviral basal-level transcription is achieved by virtue of cellular transcription factors, importantly NFκB and Sp1, Tat is made and goes to work. The establishment of viral infection can be regarded to have been bootstrapped. By binding to the TAR (transactivation target) structural element in the nascent leader RNA, Tat positively governs HIV gene expression through an increase in viral transcription or an increase in transcriptional elongation, or through a coupling of these. This process is not entirely clear: in its possible transcriptional elongation role, Tat may function like prokaryotic antitermination regulatory elements, a concept developed by Chao and colleagues. A minor protein fused of Tat, envelope, and Rev fragments, called Tev, has been reported for some viruses.[64] Cellular factors are required for Tat function and possibly for Tat downmodulation,[26] since some process must come into play to prevent the positive feedback from running wild.

All lentiviruses are known to have their own transcriptional activators, typically small proteins with an activation domain and a nuclear-localization/nucleotide-binding domain, but only some *trans*-activators have the ability to bind to a specific RNA site or interact with a TAR-like element in the nascent mRNA. Tat-like proteins are probably homologs of cellular transacting factors. Because HIV Tat has some enhancing effects on DNA elements upstream from TAR (in conjunction with the cellular factor Sp1), and in neural cells employs a TAR-independent mechanism for infection,[96] it may be considered a pleiotropic regulatory protein that has been selectively adapted to a TAR-dependent role in certain host cells, namely those that it most easily infects. Consistent with this interpretation are the findings that Tat activates host cytokine genes and is a secreted growth factor.[23,26,74] Under some circumstances it is observed to protect cells from apoptosis, or programmed cell death.[98] There are probably effects of Tat that contribute to HIV pathogenesis, however the existence and function of this protein in nonpathogenic forms of SIV must be kept in mind (although there is some speculation that viral latency may come about through Tat-defective mutants). Tat can transactivate heterologous viral promoters, for example that of papillomavirus, however it is unlikely that the two viruses will be naturally found in the same cells; we will return to this fact in our discussion of papillomavirus transcriptional control. Viral cross-talk may occur in nature between HIV and herpes virus 6, since both inhabit CD4 cells and have the ability to transactivate one another.[74]

The TAR RNA sequence, approximately 60 nucleotides in length (Figure 1), is the principal target sequence for Tat. It is thought to position, or "tether," Tat for proper interaction with the transcription complex. As we shall discuss later in this introduction, TAR may have been appropriated from cellular material, giving

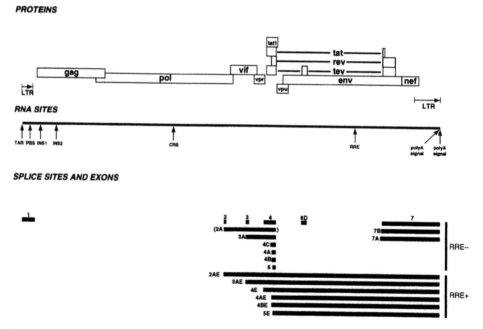

FIGURE 1 Schematic summary of HIV genome and its landmarks. Coding regions and exonic and intronic segments are shown; the number of possible transcripts made from these will be 20-30,[64] with some containing RRE, others not. The RRE, TAR, and INS elements are discussed in the text. (CRS is an INS, a *cis*-acting repressive element that acts in the absence of Rev.)

further support to the speculation that Tat's repertoire of activities has been expanded through evolutionary fabrication. TAR DNA, about which little tends to get said, also appears to serve as an transcriptional initiator element in HIV.[73] Surrounding TAR DNA in the LTR (long terminal repeat) of HIV are numerous transcriptional enhancer and promoter elements that interact with cellular factors.[42,44] A transcription termination signal is also embedded in this rich 5' regulatory region, but the nearby major splice donor site is thought sufficient to ensure that modification (polyadenylation) and transcription termination occur at the 3' site.[1] Returning to TAR RNA, an apical loop portion with which one or more cellular proteins interacts (Tat binds to a bulge 5' to this loop) can engage in a "kissing loop" structure with a sequence stretch hundreds of nucleotides downstream in the HIV Gag coding region, known as TAR∗ [7] (H&K); this highly specific interaction could be intramolecular or intermolecular and could be enhanced by protein binding. The functional significance of this novel, energetically favorable structure is unknown: among functions that could be affected are gene expression, translation, viral dimerization, replication, and recombination. In the absence of Tat, TAR acts as a site that inhibits translation of cell proteins, reportedly through activation of double-stranded RNA-dependent protein kinase.[80,81] The structure of TAR RNA and its interactions with Tat and cellular factors are described in further detail by Chao et al. and by Huynen and Konings.

In light of the intracacies of the Tat and TAR systems, meshing as they obviously do with a plethora of cellular factors and effects, one might wonder what additional complexity can be layered upon the "single-minded" retrovirus. A second major regulatory scheme in HIV is the Rev-RRE system. To understand the logic of this system, we should take note of the many transcripts that are produced by HIV[57,64] (Figure 1). Alternative splicing is a prominent regulatory feature of many small viruses. In addition to a major splice site found in retroviruses in general—which provides a mechanism to produce the additional envelope transcript from the primary viral RNA transcript needed for Gag and Pol—there are numerous (twenty-to-thirty?) singly and multiply spliced messenger RNAs, which underline the potential for temporal regulation of viral gene expression.

Tat, along with some regulatory proteins still to be introduced, is transcribed from one class of the multiply spliced RNAs that are efficiently expressed in the absence of a protein known as Rev. These transcripts are characteristically devoid of potential introns as well as an RNA structure, RRE (Rev responsive element), which is located in a conserved region of the envelope gene. Rev exerts its action through RRE, therefore mRNAs without the latter are said to be Rev-independent. These are the earliest class of transcripts that become successfully translated in the cytoplasm of the infected cell. Also among the first wave of transcripts are species that encode Rev. As Rev builds up and "shuttles" between the cytoplasm and the nucleolus, where it predominantly resides, it facilitates the nucleocytoplasmic transport of a second class of transcripts, which contain RRE and therefore are Rev-dependent (Figure 1). The teleonomy of this posttranscriptional regulatory process is elegantly presented by Cullen: apparently transcripts with introns are held and/or degraded in the nucleus by so-called "commitment factors" unless rescued by a sequence-specific nuclear export factor such as Rev. As Rev and RRE come into play, Gag and envelope mRNA levels rise in the cytoplasm and now less necessary Tat and Rev transcripts decline. The Rev-dependent wave of transcripts also accounts for some accessory proteins that we will discuss below. Curiously, among these are an intermediate transcript for envelope, perhaps because the envelope must be coordinately expressed with one of the accessory proteins known as Vpu.[64]

Tat and TAR are the dominant virally expressed regulatory elements in the earliest phase of viral residency—they "jump-start" viral expression. Rev and RRE are essential factors for stability and transport of unspliced and partially spliced messenger RNAs, which in turn bring about a temporal regulation of viral expression. Rev is a viral protein that is thought to have a cellular analog (or homolog), partly because host genes are also engaged in temporal regulation through post-transcriptional control and partly because simpler retroviruses have an RRE analog but do not encode a Rev analog. Alternative splicing and control of nuclear export are extremely pervasive themes among retroviruses. For example, oncoretroviruses, such as HTLV, have a structural analog to Rev known as Rex; the corresponding RNA effector for Rex is denoted RxRE, which is found in the retroviral LTR region rather than in the envelope coding region. Although they have little sequence similarity, and are not found in homologous genomic sites, Rex-RxRE and Rev-RRE can uncannily replace one another in various "mix-and-match" experiments, and RRE-like elements (CTE, or constitutive transport elements) from simpler retroviruses can replace RRE for HIV (Felber).

The RRE RNA element is larger than the TAR structure, though the Rev-binding site appears to be a relatively small segment of this overall structure (H&K, Cullen). Rev is known to antagonize less well-defined instability regions in the viral transcripts that prevent expression of unspliced mRNAs, either by shortening their half-life or by affecting their splicing through interaction with nuclear splicing factors. These sequences, termed INS sequences (or CRS), are known to be embedded in several locations of the HIV genomic RNA: they may be binding sites for cellular splicing and/or commitment factors. An INS element has been reported in the HIV *pol* gene that may coincide with a well-defined RNA structure whose function has not been assigned[31] (H&K). Mutation of an INS sequence found in the *gag* gene has been shown to free up the *gag* need for Rev.[78] Rev is not improving the expression of stable, well-expressed mRNAs, rather it only works on problematic mRNAs. When RRE is transferred into cellular mRNAs, INS elements must also be present to make these genes Rev-responsive.[54] RRE and its ancestral analog CTE may be modified INS elements; alternatively (for reasons that become clear below), RRE may have been imported into, rather than adapted from, the viral molecule.

Four so-called accessory proteins are encoded by HIV, one of which, Nef, comes from a Rev-independent transcript and therefore is an early protein. The remaining three proteins that are produced later are Vif, Vpr, and Vpu. The roles of these proteins *in vivo* are regulatory in the broad sense, but because they are not required for growth in many cell lines, they have been separately classified from Tat and Rev. To begin to understand some of the need for these proteins, it is helpful to consider the complications posed for HIV by the CD4 molecule. The strong interaction between the HIV envelope and the CD4 molecule can take place intracellularly as well as at the cell surface prior to viral entry, thus it becomes necessary that the nascent intracellular envelopes be freed from such complexes. Further, downregulation of surface CD4 in the infected cell is seemingly desirable, as if superinfection may not be to the advantage of the provirus and its progeny. *In vitro*, Vpu is involved in intracellular CD4 degradation and Nef is associated with cell surface CD4 downregulation. Because the reading frames of these proteins subtend the HIV envelope coding sequence (Figure 1), there is some suspicion that they are envelope-derived functions; i.e., to be found on the envelopes of other retroviruses rather than as independently translated proteins.[64]

As we previously noted, the Vpu and envelope proteins are translated from the same transcript. In addition, Vpu is not found in SIVs and HIV-2 where the envelopes of these viruses perform Vpu functions. In HIV-1, Vpu also enhances release of virions from cells through mechanisms that are not well understood at this time.[40,84] Nef is an even more complex protein whose *in vivo* role is not clear,[46] though it appears to be needed early in the viral life cycle. In addition to shutting down surface CD4, it downregulates MHC class I molecules, thereby preventing thereby the recognition of the infected cell by cytotoxic T-lymphocytes.[25] Nef is separately thought to be involved in viral infectivity and also with signal transduction pathways. Within the lentiviruses, the *nef* gene is found only in the primate immunodeficiency viruses, thus it does not seem to be needed for support of the processes served by lentiviral Tat and Rev proteins. On the basis of *in vitro* studies, Nef was first considered to be a negative regulatory factor (see Neuveut and Jeang, hereafter N&J), but recent animal studies have shown that it is important for viral replication *in vivo* and contributes to viral pathogenesis. For these reasons, *nef*-defective viruses are being considered for use in attenuated vaccines.[17]

Vpr has become an intensely studied protein since the discovery of its involvement with cell-cycle regulation. HIV-infected cell cultures with mutated *vpr* genes tend to be long lived, whereas cell death rapidly ensues with intact *vpr* genes, hence the gene is sometimes called *rap*, for rapid. Vpr negative viruses appear to be at a disadvantage only in macrophages (not in lymphocytes), suggesting that the protein provides a viral function in nondividing cells that is readily available in activated cells. While Vpr stimulates viral replication, it can potently arrest cell growth in the G2 phase of the cell cycle, apparently by inhibiting cellular kinases[70] or through interaction with a protein called RIP[72]; programmed cell death can follow.[83] Vpr also assists the retroviral Gag protein with the transport of the retroviral preintegration complex into the nucleus. Because it is incorporated into virions (through the mediation of Gag), and because it is produced by both Rev-dependent and Rev-independent mRNAs,[79] it probably plays an essential role in the earliest phases of viral infection. Like the Tat protein, Vpr has LTR connected transcriptional activation properties[94] and can also exert some of its effects extracellularly. In HIV-2s and SIVs, the nuclear transport function of HIV-1 Vpr is carried out by a duplicated gene product called Vpx, whereas the cell cycle arrest activity is controlled by the HIV-2/SIV Vpr.[24] (This interesting evolutionary construction is discussed below.) Variation in the cell cycle arrest function of Vpr proteins has apparently been driven by adaptation to the specific hosts of these viruses.[85]

The last HIV accessory protein that we will mention is Vif, which is said to be a viral infectivity factor. Vif, in some instances, can be replaced by cellular factors, but it is essential for productive infection of restrictive cells.[64,71] It is now recognized to be virion-associated and, therefore, could be construed as important for early infection of both lymphocytes and macrophages.[35,45] On the other hand, there are abundant signs that its functions may be related to later stages of the viral life cycle—assembly or processing—in which case Vif may not be a regulatory element in the strong sense: for instance, it is reported to associate with a cytoskeletal filament protein, vimentin.[35]

Table 1 summarizes HIV regulatory proteins and elements and the Vif protein is included, although it doesn't appear at this time to be a regulatory protein in the strong sense. There may be additional features of HIV regulation that are still to be elucidated; however, given the enormous attention focused over the last decade on Tat, Rev, Vpr, and RNAs, *in vitro* and *in vivo*, the chances that principal regulatory elements have gone undiscovered are small. Certainly many challenging questions remain: for example, What is the precise mechanism by which Tat operates (Chao et al.)? How is the directionality of Rev controlled (Cullen)? Space doesn't permit a discussion of the many cellular molecules that are cofactors of HIV infection. For a summary and discussion of these cofactors, see Jeang.[33]

To summarize HIV regulation, this small retrovirus is capable of infecting nondividing as well as actively dividing cells through a battery of complex mechanisms. These viral processes are governed by a relatively small number of viral proteins and RNAs, some of which are multifaceted, which interact with a larger array of equally multifaceted cellular factors. Viral entry signals the start of a bootstrapping operation in which cellular transcription factors are recruited, thereby enabling the virus to break what is called molecular latency. But because the HIV genome is compact, operating under a single promoter, problems are posed for temporal expression of viral regulatory features: different classes of spliced transcripts must be marshalled by a viral nucleocytoplasmic transport mechanism, which may additionally serve to modulate viral transcription. The major elements of HIV regulation appear to be based on, and probably evolutionarily derived from, cellular regulatory strategies.

TABLE 1 Principal HIV Regulatory Proteins and Elements

Name	Size	Functions	Localization	Reference
Tat	14/16 kDa	Transcriptional activation; many other functions.	Primarily nucleus; can be extracellular	Chao et al.
Rev	19 kDa	RNA transport and stability	Primarily nucleolar, but shuttles between nucleus and cytoplasm	Cullen
Vpr	10–15 kDa	Arrests cells in G2/M; supports nuclear localization of preintegration complex; transcriptional activator	Nuclear membrane; virion; extracellular	Intro.[24,70,83] Intro.[85]
Vpx	12–16 kDa	Nuclear localization of preintegration complex	Nuclear membrane; virion	Intro.[24,85]
Vpu	16 kDa	Extracellular release of virus; degradation of CD4	Integral membrane	Intro.[40,84]
Nef	25/27 kDa	CD4 and MHC-I downregulation; other functions?	Plasma membrane; virion?	Intro.[46]
Vif	23 kDa	Virion maturation and infectivity	Cytoplasm; virion	Intro.[35,71]
TAR	60–120 nt	As RNA, tethers Tat; as DNA, promotes transcription; inhibits translation; kissing hairpin with TAR*	R region of LTR	Chao et al., Huynen and Konings
RRE	~250 nt	Rev-responsive element	Within envelope coding sequence	Cullen, Felber, Huynen and Konings
INS	~200 nt?	RNA instability; interacts with splicing/commitment factor?	Scattered throughout genome	Intro.[31,54,78]
promoter, enhancer, initiator, terminator sequences	Variable, but usually short	Transcriptional initiation and termination	5′ and 3′ LTRs	Intro.[26,33,73]

2. ORIGINS AND VARIATION OF HIV REGULATORY ELEMENTS

With complex retroviruses, it is easy to imagine that their plasticity and genomic expansion result from heterologous recombination, and there is evidence to support this hypothesis from the simpler retroviruses. The *src* gene, for example, appropriated by the *gag-pol-env*-complex in Rous sarcoma virus brings a cell-transforming capability to a previously minimal retrovirus.[3] The function of cellular homologs of retroviral oncogenes such as *sis* and *jun* are sometimes known, though many remain unknown. Viral Src has diverse functions that may relate to other proteins, for example Jun, whose genes have been independently captured by retroviruses.[16] Because the HIV *nef* gene is found at the end of the viral genome in only primate immunodeficiency viruses, it is tempting to think that it is a recently acquired element; the observation that Nef protein has phosphorylating capabilities and may be involved in signal transduction has fueled this speculation.[100] Indeed, many homology claims have been made about Nef, so many in fact that the issue is as cloudy now as when Nef was first discovered.

The picture may be somewhat clearer with the central region regulatory elements of HIV (Figure 1). Tat and Vpr are weakly similar at the protein level and, as noted above, both have transcriptional transactivation capabilities. Hence regulators of some lentiviruses may be homologs of Vpr instead of Tat. A plausible hypothesis is that an ancestral cellular gene was imported (into lentiviruses?) and duplicated to create the two functions. The precedent for this kind of evolutionary event is argued from the *vpr-vpx* homology, or, strictly speaking, parology. We recall that in HIV-1s, the single Vpr protein performs at least two major functions, whereas in HIV-2s and related SIVs, the Vpr performs one and the Vpx performs the other. These genes and their products can be aligned to one another and analyzed by phylogenetic tree analysis.[90] Although regulatory properties, as such, are not involved, another example of what appears to be a parologous relationship is observed between certain retroviral protease and dUTPase genes.[50] We have also considered previously the source of the *vpu* gene to be the adjacent *env* gene.[64]

The origins of RNA structures such as TAR and RRE may also entail heterologous recombination followed by gene duplication. Our greatest insight comes from the TAR element, which in HIV-1s is about 60 nucleotides in length, but in HIV-2 and some SIVs is about 120 nucleotides and consists of a double structure.[4,6,34] That these two RNA sequences are homologous can be shown through pairwise sequence comparison, although gaps in the alignment must be introduced and the divergence is significant. HIV-1 Tat can transactivate the HIV-2 TAR (though not vice versa).[6] What is curious, but typically unrecognized, is that the HIV-2 TAR is highly similar to a coding sequence from a variable heavy-chain immunoglobulin gene, as if TAR's source is a cellular exon with a regulatory structure embedded (Figure 2). The chance that the similarity arises through convergence (parallelism) is quite small.

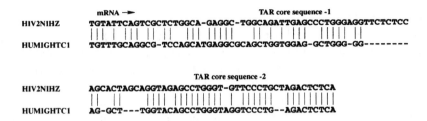

FIGURE 2   Sequence similarity between the HIV-2 TAR segment and a human variable heavy-chain immunoglobulin coding region. The mRNA start site and the duplicated TAR core regions are indicated. Gaps in the alignment are shown by dashes.

A duplicated TAR structure has been reported for one HIV-1 variant, but this seems to have arisen through homologous, rather than heterologous, recombination. The added stretch in the HIV-1 variant, while conducive to a second TAR fold, is not similar to the "duplicated" TAR seen in all HIV-2s and their relative SIVs.[22] In contrast to most HIV-2s, the HIV-1 variant has a high replicative capacity and is cytopathogenic.

An additional clue to the acquired nature of the TAR sequence comes from its base composition: HIV has a high adenine (A), low cytosine (C) makeup that is not present in TAR, or for that matter RRE; both RNA structures select for a higher G-C content.[57] The redundancy of TAR-2 results in a preference for the first TAR loop, with assistance from the second.[6] To our knowledge, no one has proposed an origin for the HIV RRE (see H&K). We speculated earlier about its possible homology with INS sites, which interact with splicing and/or commitment factors. The CTE structure in some retroviruses does not appear to be a sequence or structure homolog of RRE but rather an independently evolved element with analogous function (Felber).

Because the pathogenesis of HIV-2 infection is milder in general than that of HIV-1 infection in humans, researchers have sought to explain these differences on the basis of TAR and LTR sequence differences.[6,34] But again, viral replication dynamics may not directly map into pathogenesis. Indeed, some long-term survivors of HIV-1 infection have mutant regulatory elements or proteins[18] and different mRNA transcript profiles can be observed in different AIDS patients[64]; nevertheless, all of the fundamental regulatory elements summarized in Table 1 can be fully present in animals that are free of disease.[38,75] This is not to imply that HIV is not the cause of AIDS; rather, it suggests that there are many ways to be an AIDS virus and that the general outcome of infection can be complex dysregulation of cell function. We will now turn to the general phenomenon of HIV variability.

RNA viruses such as HIV haven't the error-correcting capabilities of eukaryotic cells or DNA viruses such as papillomavirus, hence their mutation rate can be as high as 1 base per genome per replicative cycle. This rate is thought to be close to a proposed "error threshold" beyond which sequences become random, information is lost, and the viral population cannot survive.[61] As it turns out, a large number of viruses will be defective, or at least will not succeed; the issue here has to do with the population survival. Small RNA viruses, in general, mutate at these high rates, giving rise to swarms of viruses often called "quasispecies"; the term was originally applied to the earliest rapidly evolving populations of life on earth but has been increasingly applied to viral populations today.[30] To understand the extraordinariness of HIV, however, it is important to distinguish mere replicative errors (and recombination) from variation, the result of mutation and forces of selection.[12] HIV appears to be one of the most variable viruses known, demonstrating a different pattern of evolutionary diversification than, say, influenza A virus, which also mutates rapidly: influenza viruses display strong competition with one another, as evidenced by their evolutionary pattern of sequential replacement, whereas HIVs show little propensity for such replacement. HIV variation is further explained by considering average ratios of nonsynonymous to synonymous substitutions: the former measure rates of amino acid-replacing substitutions whereas the latter do not result in amino acid changes. Many RNA viruses, for example hepatitis C virus and polio, mutate rapidly but display lower ratios of nonsynonymous to synonymous substitutions. Importantly, nonpathogenic SIVs from animals in the wild possess lower ratios than what is seen with HIV-1.

When it was first noted that HIV varied rapidly, few genes and proteins were analyzed and the generalization emerged that HIV regulatory and accessory genes were relatively conserved compared to the viral envelope, which varies under pressure from the immune system. As was noted earlier, HIV envelope is indeed variable, however so are the Tat, Rev, and Nef proteins. Turning our focus from rates to extents of variation, an appropriate information measure can be employed to show that the average information densities of Tat, Rev, and Nef are as low as envelope or any HIV protein.[37,57] To provide the reader with some sense of the breadth of Tat protein variability, a phenogram constructed from diverse amino acid sequences taken from the HIV sequence database is shown in Figure 3. A cladistic analysis of HIV-1 sequences reveals the existence of at least ten distinct sequence subtypes,[59] and probably no two AIDS patients in the world have identical viruses.

Should we be surprised at the number of ways to be Tat? Schuster and Stadler argue that many sequences tend to map over into nearly equivalent ("near neutral") structures, and therefore that redundancy is to be expected in nature. One might think that natural selection would quickly narrow the field of possible Tat proteins to the very fittest, but Wain-Hobson (this volume), following a similar line of argument (albeit with a different form of argument), gives new perspective to this assumption.

The essential domains of HIV's regulatory elements are more easily defined and manipulated as a consequence of the large rate and extent of variation. Less variable viruses do not reveal themselves much as HIV. And there are concomitant possibilities for drug discovery through combinatoric experiments using the highly variable HIV molecules (Frankel and Ellington).

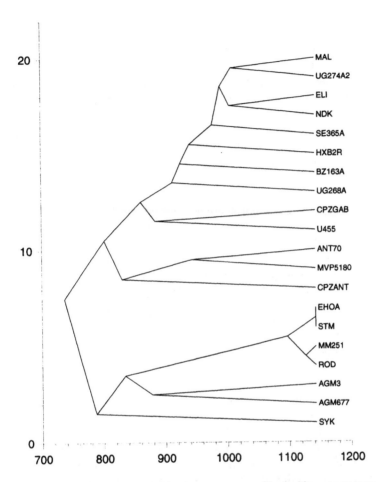

FIGURE 3 Cluster analysis of Tat protein sequences. Similarities, as measured by a BLOSUM matrix that takes into account chemical equivalences, are scored on the abscissa; the ordinate merely records the number of sequences analyzed. The extent of variation, as measured by the depth of the root node, is comparable to what is seen with the most heterogeneous proteins of HIV.

3. PAPILLOMAVIRUSES

Although researchers have been working with papillomaviruses (PVs) since their discovery in cottontail rabbits in 1933, we actually know more about HIV, which was discovered fifty years later, simply because the latter has been more easily grown and studied in the lab. More than 12,000 citations concerning HIV have been entered into the molecular subset of Medline at the National Library of Medicine in the past decade, when fewer than 1700 were recorded for papillomaviruses; yet it can be said that PVs are reasonably well-characterized viruses. In the absence of a complete culturing potential, organotypic, or epithelial raft, cultures have proven useful for study of late viral functions in certain of the HPV types[8,53] (Thierry and Yaniv, hereafter T&Y). Cloning and DNA sequencing have provided an additional impetus for molecular investigations. Of the PVs, the bovine papillomavirus, BPV-1, has been the most comprehensively studied virus at the molecular level because its early genes can be expressed and its DNA can be replicated in cultured rodent cells (though viral progeny are not produced).

PVs are small, double-stranded DNA tumor viruses related to polyomaviruses and vacuolating viruses, of which simian virus 40 (SV40) is the prototype. Virions of the three genera possess circular genomes, but no envelope, and their capsid proteins are uniformly icosahedral. Together these viruses constitute the papovaviridae family. While all three viruses are found in humans, where they can be associated with diseases, PV infections appear to be the most widespread. Incidence of human papillomavirus (HPV) transmission is as high as that for any sexually transmitted pathogen, and both transmission and disease are facilitated by HIV-caused immunosuppression. It is now known that papovaviruses can be harbored subclinically, that is to say with little or no adverse effects. Most of the diverse forms of PVs found in reptiles, birds, various mammals, and primates have become adapted to their hosts and connected with a variety of benign skin lesions—papillomas, or warts.[5,67,91] The tropism for PVs encompasses cutaneous and mucosal epithelium, and the permissivity is said to be strict.

Certain types of human papillomaviruses, which are said to be "high-risk," can be causative agents in cervical and other anogenital cancers; cervical cancer is a leading cause of cancer deaths in women worldwide.[51] There are additional tumor-causing species of HPV associated with rare skin cancers.[67,82] HPV-related oncogenesis should be regarded as an anomaly of persistent infection since it is not associated with viral production.[52] We will return to this fact when we examine transformation and its factors in order to better understand the benign as well as the malignant state. One informative type of study contrasts mucosal infections by high-risk viruses which, in rare cases and in concert with other factors, result in malignancy, to infections by low-risk viruses, for which the clinical prognosis is limited to genital warts (condylomata).

The life cycle of papillomaviruses is played out within layers of differentiating epithelial tissue in which the virus can be viewed as holding in check its activities in order to conform to the cellular program (Figure 1 in McBride et al.). PVs invade the basal (or innermost) layer that contains a small number of stem cells, probably through a lesion in the epithelium. Infection is thought to begin with invasion of these slowly proliferating cells rather than of the nondividing cells.[8] In any case, as the cells move outward (upward) toward the surface, they normally differentiate and do not produce factors conducive to DNA replication. PV early proteins (and perhaps the wound that has allowed the virus to gain entry in the first place) redirect this process by stimulating cell proliferation, thereby promoting viral DNA replication. Nevertheless, differentiation of the epithelial layers continues in most cases, hence modulation rather than total redirection of host-cell processes is the outcome. Viral production may contribute to apoptosis, but the differentiated keratinocytes approaching the skin surface (stratum corneum) are already destined to death.

PVs accomplish their reproductive goal as intact extrachromosomal plasmids (episomes): approximately 20–100 copies are found per cell in the basal layer of cutaneous infections and will increase to thousands of copies in the upper spinous strata.[53] A rare accident in this cycle leads to viral integration into human chromosomal DNA, with key regulatory elements of the virus becoming lost in the integration process. In this event, the upper layers of cells tend to remain undifferentiated, characteristic of a high-grade carcinoma, and no virus is produced.

We will introduce the HPV proteins and regulatory structures in the approximate order they are transcribed (Figure 1 in T&Y, Figure 1 in Baker): The early region proteins E6 and E7 may first act by influencing basal cell proliferation, however they are most evident in the suprabasal stage of infection.[8] E2 protein is essential from the outset for viral transcription and E1 for viral replication; these latter proteins work together to maintain viral episomes. An early protein named E5 probably also contributes to induction of stem cell growth. A sixth early region protein, E4, is actually made late and seems to have a mostly architectural role. Finally, capsid proteins L1 and L2 are temporally expressed such that viral encapsidation coincides with terminal differentiation of the squamous epithelium. These proteins will be briefly described at this time as they are superbly covered by McBride, Androphy, and Munger.

The papillomavirus E6 protein is a remarkable multifaceted protein. It is known to be a transcriptional activator, a modulator of cell-cycle processes, a cofactor in the degradation of the tumor suppressor protein p53, and a transforming agent capable of immortalizing cells in culture. Yet, E6 is present in very low concentrations in the infected cell and has not been as well-characterized as, say, the HIV Tat protein; it has been reported to bind numerous cellular factors. E6 functions as a transcriptional activator for both high-risk and low-risk viruses, perhaps through interaction with a still to be characterized regulatory sequence in the viral long control region (LCR),[41,62] see T&Y, Figure 2. More importantly, this relatively small protein may be a heterologous *trans*-activator or a transcriptional repressor; it possesses two large zinc domains characteristic of a certain class of cellular transcription factors,[14] and it binds numerous cell factors. No cellular homolog of E6 has been identified. While the E6 transforming properties in high-risk viruses could be coupled to their transactivation properties, E6s from the majority of PVs suggest otherwise in so far as they transactivate but do not transform.[2] (Some PVs, for example BPV-3, BPV-4, etc., do not have an E6 gene.[65])

The cell protein p53 is an important tumor suppressor protein (due to its transcriptional activation properties) whose degradation is targeted by the E6 of high-risk viruses.[97] These E6s bind a cellular protein named E6AP to participate in a ubiquitination pathway that accomplishes selective degradation of p53.[8,77] E6 may also repress p53 transcription. As pointed out by McBride and colleagues, one of the primary roles of E6 in high-risk viruses is to prevent p53 from arresting cell growth. But E6 proteins from low-risk viruses do not target p53 degradation, thus it is not clear whether the general biological function of E6 for the majority of PVs concerns p53. Neuvelt and Jeang present several examples of confounding observations in the field of viral biology. In this context, to make the generalization that E6 and E7 primarily act through binding tumor suppressor proteins could be to overinterpret what has been learned from the greater ability to characterize high-risk viruses. It is significant, nonetheless, that regulatory proteins of other DNA tumor viruses, such as adenovirus, have strategies for antagonizing growth control; in the adenovirus case through sequestration of p53.[77]

The activities of another oncoprotein, E7, are closely coordinated with E6. E7 is also a zinc finger protein with transcriptional activation properties; in particular, it interacts with E2F, a prominent cellular transcription factor and common target of tumor viruses,[15] and retinoblastoma protein (Rb), another tumor suppressor protein. These interactions drive normally quiesent cells into S (synthesis) phase, which is needed for viral DNA replication. E7 performs several other roles that are also seen in the adenovirus E1A protein and the SV40 large T antigen, both of which show sequence similarities with the papillomavirus E7.[8,77] p53 transcription is reported to be affected by E7,[48] and the transforming properties of E7 and E6 complement one another, with the largest effects encountered when both proteins are present.

The synthesis of E6 and E7 are under the control of the viral E2 protein, a transcriptional activator whose target-enhancer sequences reside in the LCR (or URR), a noncoding region of about 800 nucleotides. Ten or more cellular proteins are also known to bind to the PV LCR region, some of which are involved in jumpstarting HIV (T&Y). E2 binding sites in the LCR vary in number from 4 to 17, depending upon the PV type, and their position becomes important[62]: E2 at high concentrations has repressor activity, thus the current hypothesis is that at low concentrations, full-length E2 binds to high-affinity upstream sites in support of activation, while at high concentrations it binds to lower affinity sites that may interfere with TATA and Sp1 promoter elements. The transactivation domain of E2 is located at the N-terminal end of the protein; smaller proteins made from the 3' portion of the E2 orf function as transcriptional repressors, possibly by competing for high affinity E2 binding sites. Because E2 can repress E6 transcription and E6-mediated immortalization of cells, it can be viewed as supportive of p53 function in high-risk infections.

The full-length E2 protein is centrally involved in viral replication, although the principal molecule responsible for initiating PV replication is E1, a large viral protein that has local similarities to domains of the large T antigen of SV40 and polyomavirus.[74] E1 can bind to DNA nonspecifically, but it binds to a specific origin of replication (ori) site when in complex with E2.[86] E1 reciprocally affects E2 in its role as a transcriptional activator/repressor, thereby playing an indirect role in immortalization processes. The overall effect of E1 is one of negative regulation of oncogene expression.[86] Integrated HPVs from tumors are often found to have disrupted E1-E2 genomic regions, accounting for the absence of controls over E6 and E7 and for the inability of these tumor cells to support viral production[8] (McBride et al.). An additional consequence of integration is that integrated viruses are unable to synthesize the L1 and L2 capsid proteins, as the coding regions for these are now dissociated from their promoter.

HPV E5 is a transforming protein encoded by a gene immediately downstream from the E2 coding region. It appears to participate in a wide variety of activities, but is best characterized for its ability to transform rodent fibroblasts. Levels of cellular oncogenes, *c-fos* and *c-jun*, as well as cell growth factors such as MAP kinase, are upregulated by E5.[27] The picture with E5 is partly unclear because different, perhaps nonhomologous, proteins are equivocally named E5 (see below). Some HPV E5s and BPV-1 E5 are known to form a complex with a 16 kiloDalton protein that is a subunit of vacuolar H^+-ATPase, tending to abrogate normal 16K functions, which are acidification of endosomes and mediation of receptor-ligand interactions. In this role, BPV-1 E5 displays functional similarities to other viral proteins, in particular the HTLV-I p12^I, and there are some limited sequence similarities between these apparently convergent proteins.[27] The BPV-1 E5 also binds to the PDGF (platelet-derived growth factor) receptor.

The E4 protein is also extremely divergent, making it difficult to pin down its functions. In cutaneous HPV-1 infections, relatively large amounts of E4 are made and the level of expression is correlated with that of late proteins, L1 and L2. Hence, E4 may be a structural protein associated primarily with keratin filaments, and it may be synthesized late rather than early in the viral life cycle. The L1 and L2 proteins are the major and minor capsid proteins, respectively. E4, L1, and L2 are not included in Table 2, which summarizes the PV regulatory elements.

Papillomaviruses, like HIV, manufacture multiple mRNA transcripts (T&Y, Baker) and utilize complicated splicing to regulate the expression of these; unlike HIV, they have to regulate early and late promoters. Balanced splicing and alternative polyadenylation are some of the intricate features of temporal expression of PV transcripts: we saw in HIV that a splice donor site helps regulate polyadenylation,[1] and a similar interaction occurs with HPV splicing and polyadenylation (see Baker). The RNA structures required for these processes can be located in some HPVs, however it has not been possible to identify "well-defined structures" common to all PVs, or to all PVs of a given class (H&K). This inhomogeneity of RNA structures, in contrast to the conservation of HIV structures such as TAR and RRE, is somewhat puzzling. A 3′ GU-rich negative regulatory element interacts with cellular proteins (commitment factors?) to antagonize expression of late mRNA in undifferentiated cells.[19] In HPV-1, an inhibitory sequence similar to what is encountered in HIVs, an AU-rich segment in the 3′ untranslated region of late mRNAs, interacts with cellular proteins. It is noteworthy that the inhibition it imposes on late mRNA expression in undifferentiated cells can be overcome by introduction of a CTE (Felber) or the Rev-RRE system (Cullen).[88]

TABLE 2 Principal HPV Regulatory Proteins and Elements

Name	Size	Functions	Localization	Reference
E2	h 48 kDa (full-length); smaller proteins encoded by 3′ portion of E2 orf.	Transcriptional activator and repressor; supports E1 in viral replication; packaging; other.	nuclear and cytoplasmic	McBride et al., Thierry and Yaniv
E1	~70 kDa; E1-M is 23 kDa	Initiator of viral replication; modulates E2.	nuclear	McBride et al., Thierry and Yaniv
E6	~18 kDa	Multifunctional; transcriptional transactivator; transforming protein; targets p53 degradation; binds cellular proteins, e.g., E6BP.	nuclear and cytoplasmic	McBride et al.
E7	10–14 kDa	Multifunctional; transcriptional transactivator; transforming protein; binds Rb and dissociates Rb-E2F complexes; other.	nucleolar	McBride et al.
E5	~7 kDa (in BPV-1)	Transforming protein; binds PDGF receptor and 16 kDa ATPase subunit.	membranes	Intro.[8,27]
LCR	~800 nt (in genital HPVs)	Contains *cis*-responsive elements involved in transcription and origin of replication.	5′ terminus of linear genome	Thierry and Yaniv
NRE	~80 nt	Negative regulatory element (*cis*) affecting late mRNA expression	3′ untranslated region	Baker
promoters, enhancers, splice enhancers and suppressors, terminator sequences	variable, but usually short	Transcriptional and posttranscriptional control	scattered throughout genome	Thierry and Yaniv, Baker, Huynen and Konings

4. PAPILLOMAVIRAL ORIGINS AND VARIATION

Papillomaviruses are less likely than retroviruses to acquire genetic material, first because the integrated form of PVs does not produce virus and second because there is no significant evidence that PVs engage in either homologous or heterologous recombination. They appear to be very old viruses: those PV proteins that are similar to other viral proteins probably have arisen through divergence (E7 and large T antigen, E1 and large T antigen) or convergence (E5 and HTLV-1 p12^I). It has been suggested that E6 and E7 are parologously related, however no mechanism or precedent for gene duplication has been described in papillomaviruses. It is also reasonable to hypothesize that E6 and E7 are distantly related to cellular transcription factors that have large zinc finger structures, but to our knowledge no candidate proteins have been put forward.

More than 75 types of human and animal PVs have been characterized through genetic sequencing[58,67,91] (Figure 4). When weighted parsimony is utilized for analysis of the deeper relationships between these viral types, large genetic distances are implied. Yet, some animal PVs are closely related to human PVs, which suggests that cross-species transmission (which is common with retroviruses and many other viruses) is likely.[58] (For a contrary opinion, see Bernard.[5]) DNA viruses not only mutate much more slowly than RNA viruses, the ratio of nonsynonymous to synonymous substitutions in PV coding sequences is as low as what is seen with typical host genes; as a result, amino acid replacements are relatively uncommon.

HIVs may have accumulated as much diversity in the last fifty years as PVs gathered in hundreds of thousands of years. In rare instances, however, PVs have displayed lack of linkage and saltatory genetic changes.[58,76] Further, because the extent of variation is large—the viruses have undoubtedly been circulating and evolving for a very long time—the degeneracy of PV proteins is great. Figure 5 is a fair representation of E5 protein sequence diversity: E5 protein sequences from monophyletically related PVs can differ by 50% or more, manifesting as much divergence as HIV Tat proteins (Figure 3). It is not entirely clear that all of these proteins are orthologously related. On the other hand, note that the HPV-13 and chimpanzee PCPV1 proteins are unusually similar.

As we find with diverse HIV sequences, PV protein motifs can be readily identified against the extensive backdrop of sequence variation. Thus, blocks of conserved protein sequences can be quantitatively defined for E6 and E7 proteins, leaving open the possibility that critical molecular differences in these proteins from high-risk and low-risk viruses will reside in the nonconserved regions.[58]

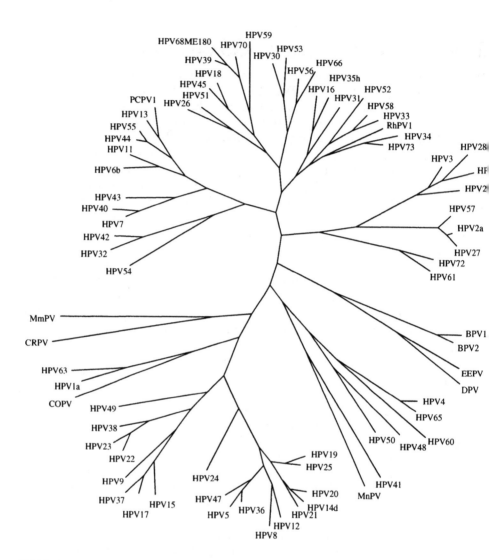

FIGURE 4 Phylogenetic tree showing the diversity of human and animal papillomaviral DNA sequences. The reconstruction is based upon the first- and second-base positions in codons of E6 coding sequences and inversely weighted parsimony.[58] The branch lengths are not proportional to single-base substitutions, but the relative lengths are highly informative. Note the close relationship between HPV-13 and a chimpanzee isolate, PCPV1.

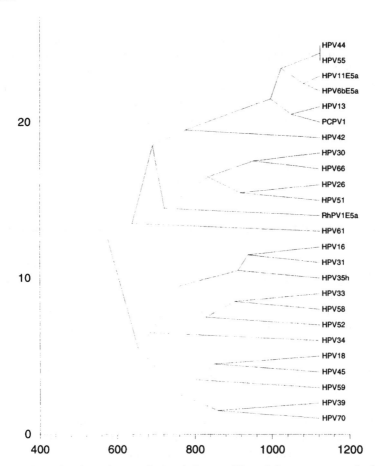

FIGURE 5 Phenetic clustering analysis of diverse E5 protein sequences. As in Figure 3, cluster analysis was performed using a BLOSUM matrix, which scores similarity as well as identity of amino acids (abscissa). Although the evolutionary rates of HPVs are much slower than the rates of HIVs, the extent of diversity is comparable to what is seen with HIV Tat (Figure 3). The ordinate merely records the number of sequences being clustered.

5. SUMMARY

The strategies that HIV and papillomaviruses have evolved to generate viral progeny are typical, and the extent of variability of their regulatory structures is not unrepresentative. Many viruses have control mechanisms for selectively activating, repressing, or modulating cellular transcription. The HTLV Tax protein has been mentioned as a multifunctional retroviral protein that displays both transcriptional activation and transforming properties. While the extent of Tax variation is significant among different HTLV and STLV types, heterologous transactivation and pleiotropic effects are common, not unlike what is seen with HPV transforming proteins.[26] To augment their complexity, HIV and HPV have invested heavily in the regulation of splicing and posttranscriptional processing of mRNAs. Vesicular stomatitis virus and influenza are two other highly degenerate viruses that have analogous strategies for regulating cellular and viral splicing.[28] Adenovirus is an example of yet another virus that possesses an elaborate array of early and late genes which are temporally controlled.[68]

Viruses also embody mechanisms of translational control. It is not at all clear what the precise purpose of the HIV TAR segment is in this regard, but it is observed to have an inhibitory effect on cellular protein synthesis.[80,81] Vaccinia virus and reovirus accomplish a similar outcome utilizing viral proteins rather than RNA. Some viruses participate in "host-cell shutoff phenomena" by interfering with the ribosome scanning process characteristic of most eukaryotic machinery.[49] Hepatitis C virus (HCV) and polio are examples of viruses that employ a cap-independent mechanism for initiating translation: rather than relying on orthodox initiation and elongation (that in the case of polio is totally blocked by the virus), a special RNA structure of about 300 nucleotides in length, known as an IRES (internal ribosome entry site), is employed for translational initiation at the 5' end of the viral genome.[32,49,93] These do not appear to be homologously related RNAs in HCV and polio, therefore they could represent separately acquired elements or parallel evolution. The translation of polio transcripts is greatly enhanced by a cellular protein, the La autoantigen, that is known to also interact with HIV TAR.[87] Perhaps if the origin of TAR can be successfully tracked, we may also gain some insight into the origin of the IRES element. We might ask whether any of the unidentified RNA structures uncovered by Huynen and Konings could be an IRES, or could be involved in unusual translation initiation events.

A prominent theme of viral infection, as we have seen, concerns the involvement of viral proteins in apoptosis, or programmed cell death. In general, viral proteins such as Tat promote cell death, giving rise to speculations that apoptotic mechanisms originally evolved as a defense against viral infection. Apoptosis is also a necessary step for differentiation and development. This suggests a possibility that the teleology of viruses may be connected in some fundamental evolutionary way to cell differentiation, an idea that has been in circulation since the discovery of endogenous retroviral elements.

The evolution of antiapoptotic viral proteins, exemplified by the *bcl-2* family found in the adenovirus E1b protein,[28] should come as no surprise: virtually every regulatory action described for eukaryotic systems has its counterflow—transcriptional activators become suppressors, structures for termination become initiators, inhibitory sequences become facilitators. One of the cellular factors that interacts with both HIV and HPV is YY1, for "yin and yang" 1. Hence, extracellular Tat protein is also reported to be antiapoptotic, perhaps because it stimulates *bcl-2*.[98] It is not to the benefit of viral replication to have programmed cell death, therefore we observe interference of cellular safeguards (e.g., the p53 system) by viral proteins (e.g., E6). When antiapoptosis becomes dominant, however—in HPVs through what appear to be mishaps in the viral-host relationship—neither differentiation nor viral replication can be fulfilled.

From our earliest awakening to the logic of gene regulation in the Lac system of E. coli, induction and repression of microbial genes could be viewed as amazingly simple negative feedback loops. Very quickly, however, complexities became apparent in the form of catabolite repression mediated by cyclic AMP, reversible constitutive mutants, and translational coupling.[44] Protein networks and cascades are now understood to delineate the web of virus-host relationships. With viral gene expression, combinations and arrangements of promoter and enhancer elements targeted by cellular and viral proteins are the rule, and we must learn how to map them over into one another. It is probably the case that no single transcription factor has primacy in any particular system. As we develop schemata for this network of regulatory elements, we are almost intellectually overwhelmed by the variety and potential. Thus, the time is ripe for innovative computational approaches such as that outlined by Somogyi in this volume. We might also find strength in the realization that when viral regulatory mechanisms are compared, we witness similarities among dissimilars, overlaps, convergences resulting from degenerate strategies. Viral specificity itself becomes comprehensible through the analysis of familiar pathways and shared purposes.

ACKNOWLEDGMENTS

We thank Carl Baker for his critical reading of the section of this chapter concerned with papillomaviruses. We are also grateful to Hong Lu and Charles Calef for their technical help with figures.

REFERENCES

1. Ashe, M. P., P. Griffen, W. James, and N. J. Proudfoot. "Poly (A) Site Selection in the HIV-1 Provirus: Inhibition of Promoter-Proximal Polyadenylation by the Downstream Major Splice Donor Site." *Genes & Devel.* **9** (1995): 3008–3025.

2. Barbosa, M. S., W. C. Vass, D. R. Lowy, and J. T. Schiller. "*In Vitro* Biological Activities of the E6 and E7 Genes Vary Among Human Papillomaviruses of Different Oncogenic Potential." *J. Virol.* **65** (1991): 292–298.

3. Benjamin, T., and P. K. Vogt. "Cell Transformation by Viruses." In *Virology*, edited by B. N. Fields, 317–367, 2nd ed. New York: Raven Press, 1990.

4. Berkhout, B. "Structural Features in TAR RNA of Human and Simian Immunodeficiency Viruses: A Phylogenetic Analysis." *Nucleic Acids Res.* **20** (1991): 27–31.

5. Bernard, H.-U. "Animal Papillomaviruses." In *Human Papillomaviruses 1997*, edited by G. Myers et al., III, Los Alamos, NM: Theoretical Division, Los Alamos National Laboratory, 1997.

6. Browning, C., J. M. Hilfinger, S. Rainier, V. Lin, S. Hedderwick, M. Smith, and D. M. Markovitz. "The Sequence and Structure of the 3′ Arm of the First Stem-Loop of the Human Immunodeficiency Virus Type 2 Transactivation Responsive Region Mediates Tat-2 Transactivation." *J. Virol.* **71** (1997): 8048–8055.

7. Chang, K.-Y., and I. Tinoco. "The Structure of an RNA 'Kissing' Hairpin Complex of the HIV TAR Hairpin Loop and Its Complement." *J. Mol. Biol.* **269** (1997): 52–66.

8. Chow, L., and T. R. Broker. "Small DNA Tumor Viruses." In *Viral Pathogenesis*, edited by N. Nathanson et al., 267–301. Philadelphia: Lippencott-Raven, 1997.

9. Clerici, M., and G. M. Shearer. "An Occam's Razor Approach to the Immunopathogenesis of HIV Infection." *AIDS* **9** (Suppl. A) (1995): S33–S40.

10. Coffin, J. M. "Retroviridae and Their Replication." In *Virology*, edited by B. N. Fields, 1437–1500, 2nd ed. New York: Raven Press, 1990.

11. Coffin, J. "HIV Viral Dynamics." *AIDS* **10** (Suppl. 3) (Dec. 1996): S75–S84.

12. Coffin, J. M. "Genetic Diversity and Evolution of Retroviruses." In *Genetic Diversity of RNA Viruses*, edited by J. J. Holland, 143–164. Berlin: Springer-Verlag, 1992.

13. Coffin, J., S. Hughes, and H. Varmus, eds. *Retroviruses*. Cold Spring Harbor: Cold Spring Harbor Press, 1997.

14. Coleman, J. E. "Zinc Proteins: Enzymes, Storage Proteins, Transcription Factors, and Replication Proteins." *Ann. Rev. Biochem.* **61** (1992): 897–946.

15. Cress, W. D., and J. R. Nevins. "Use of the E2F Transcription Factor by DNA Tumor Virus Regulatory Proteins." In *Transcriptional Control of Cell Growth: The E2F Gene Family*, 63–78. Berlin: Springer-Verlag, 1995.

16. Curran, T., and P. K. Vogt. "Dangerous Liaisons: Fos and Jun, Oncogenic Transcription Factors." In *Transcriptional Regulation*, edited by S. L. McKnight and K. R. Yamamoto, 797–831. Cold Spring Harbor: Cold Spring Harbor Laboratory Press, 1992.

17. Daniel, M. D., F. Kirchoff, S. C. Czajak, P. K. Sehgal, and R. C. Desrosiers. "Protective Effects of a Live Attenuated SIV Vaccine with a Deletion in the nef Gene." *Science* **258** (1992): 1938–1941.

18. Deacon, N. J., A. Tsykin, A. Solomon, K. Smith, M. Ludford-Menting, D. J. Hooker, D. A. McPhee, A. L. Greenway, A. Ellett, C. Chatfield et al. "Genomic Structure of an Attenuated Quasispecies of HIV-1 from a Blood Transfusion Donor and Recipients." *Science* **270** (1995): 988–991.

19. Dietrich-Goetz, W., I. M. Kennedy, B. Levins, M. A. Stanley, and J. B. Clements. "A Cellular 65-kDa Protein Recognizes the Negative Regulatory Element of Human Papillomavirus Late mRNA." *Proc. Natl. Acad. Sci., USA* **94** (1997): 163–168.

20. Doms, R. W., and S. C. Peiper. "Unwelcomed Guests with Master Keys: How HIV Uses Chemokine Receptors for Cellular Entry." *Virology* **235** (1997): 179–190.

21. Dragic, T., V. Litwin, G. P. Allaway, S. R. Martin, Y. Huang, K. A. Nagashima, C. Cayanan, P. J. Maddon, R. A. Koup, J. P. Moore, and W. A. Paxton. "HIV-1 Entry into CD4+ Cells Is Mediated by the Chemokine Receptor CC-CKR-5." *Nature* **381** (1996): 667–673.

22. Emiliani, S., C. Delsert, C. David, and C. Devaux. "The Long Terminal Repeat of the Human Immunodeficiency Virus Type 1 GER Isolate Shows a Duplication of the TAR Region." *AIDS Research Human Retroviruses* **10** (1994): 1751–1752.

23. Ensoli, B., L. Buonaguro, G. Barillari, V. Fiuorelli, R. Gendelman, R. A. Morgan, P. Wingfield, and R. C. Gallo. "Release, Uptake, and Effects of Extracellular Human Immunodeficiency Virus Type 1 Tat Protein on Cell Growth and Viral Transactivation." *J. Virol.* **67** (1993): 277–287.

24. Fletcher, T. M., B. Brichacek, N. Sharova, M. A. Newman, G. Stivahtis, P. M. Sharp, M. Emerman, B. H. Hahn, and M. Stevenson. "Nuclear Import and Cell Cycle Arrest Functions of the HIV-1 Vpr Protein Are Encoded by Two Separate Genes in HIV-2/SIV(SM)." *EMBO J.* **15** (1996): 6155–6165.

25. Gall S. L., M.-C. Prevost, J.-M. Heard, and O. Schwartz. "Human Immunodeficiency Virus Type 1 Nef Independently Affects Virion Incorporation of Major Histocompatibility Complex Class I Molecules and Viral Infectivity." *Virology* **229** (1997): 295–301.

26. Gaynor, R. B. "Regulation of Human Immunodeficiency Virus Type 1 Gene Expression by the Transactivator Protein Tat." In *Transacting Functions of Human Retroviruses*, edited by I. S. Y. Chen, H. Koprowski, A. Srinivasan, and P. K. Vogt, 51–78. Berlin: Springer-Verlag, 1995.

27. Halpern, A. L., and D. J. McCance. "Papillomavirus E5 Proteins." In *Human Papillomaviruses 1996*, edited by G. Myers et al., III-81–III-111. Los Alamos, NM: Los Alamos National Laboratory, 1996.

28. Hardwick, J. M., and D. E. Griffen. "Viral Effects on Cellular Functions." In *Viral Pathogenesis*, edited by N. Nathanson et al. 55–83. Philadelphia: Lippencott-Raven, 1997.

29. Holland, J. H. *Hidden Order: How Adaptation Builds Complexity*. Reading, MA: Addison-Wesley, 1995.

30. Holland, J., J. C. De La Torre, and D. A. Steinhauer. "RNA Virus Populations as Quasispecies." In *Genetic Diversity of RNA Viruses*, edited by J. J. Holland, 1–20. Berlin: Springer-Verlag, 1992.

31. Huffmann, K. M., and S. J. Arrigo. "Identification of *Cis*-Acting Repressor Activity Within Human Immunodeficiency Virus Type 1 Protease Sequences." *Virology* **234** (1997): 253–260.

32. Jackson, R. J. "A Comparative View of Initiation Site Selection Mechanisms." In *Translational Control*, edited by J. W. B. Hershey, M. B. Matthews, and N. Sonenberg, 71–112. Cold Spring Harbor: Cold Spring Harbor Laboratory Press, 1996.

33. Jeang, K.-T. "HIV and Cellular Factors." In *Human Retroviruses and AIDS 1996*, edited by G. Myers, B. Korber, B. Foley, K.-T. Jeang, J. W. Mellors, and S. Wain-Hobson, IV-1–IV-13. Los Alamos, NM: Theoretical Division, Los Alamos National Laboratory, 1996.

34. Jeang, K. T., and A. Gatignol. "Comparisons of Regulatory Features Among Primate Lentiviruses." In *Simian Immunodeficiency Virus*, edited by N. L. Letvin and R. C. Desrosiers, 123–144. Berlin: Springer-Verlag, 1994.

35. Karczewski, M. K., and K. Strebel. "Cytoskeleton Association and Virion Incorporation of the Human Immunodeficiency Virus Type 1 Vif Protein." *J. Virol.* **70** (1996): 494–507.

36. Karray, S., and M. Zouali. "Identification of the B Cell Superantigen-Binding Site of HIV-1 gp120." *Proc. Natl. Acad. Sci., USA* **94** (1997): 1356–1360.

37. Korber, T. M., E. E. Allen, A. D. Farmer, and G. L. Myers. "Heterogeneity of HIV-1 and HIV-2." *AIDS* **9**, (Suppl. A) (1995): S5–S18.

38. Kurth, R., and S. Norley. "Simian Immunodeficiency Viruses of African Green Monkeys." In *Simian Immunodeficiency Virus*, edited by N. L. Letvin and R. C. Desrosiers, 21–33. Berlin: Springer-Verlag, 1994.

39. Lackner, A. A. "Pathology of Simian Immunodeficiency Virus Induced Disease." In *Simian Immunodeficiency Virus*, edited by N. L. Letvin and R. C. Desrosiers, 36–64. Berlin: Springer-Verlag, 1994.

40. Lamb, R. A., and L. H. Pinto. "Do Vpu and Vpr of Human Immunodeficiency Virus Type 1 and NB of Influenza B Virus Have Ion Channel Activities in the Viral Life Cycle?" *Virology* **229** (1997): 1–11.

41. Lamberti, C., L. C. Morrissey, S. R. Grossman, and E. J. Androphy. "Transcriptional Activation by the Papillomavirus E6 Zinc Finger Oncoprotein." *EMBO J.* **9** (1990): 1907–1913.

42. Latchman, D. *Gene Regulation: A Eukaryotic Perspective*. London: Unwin Hyman Ltd., 1990.

43. Levy, J. *HIV and the Pathogenesis of AIDS*. Washington, DC: ASM Press, 1994.

44. Lewin, B. *Genes VI*. Oxford: Oxford University Press, 1997.

45. Liu, H., X. Wu, M. Newman, G. M. Shaw, and J. C. Kappes. "The Vif Protein of Human and Simian Immunodeficiency Viruses Is Packaged into Virions and Associates with Viral Core Structures." *J. Virol.* **69** (1995): 7630–7638.

46. Luo, T., J. L. Foster, and J. V. Garcia. "Molecular Determinants of Nef Function." *J. Biomed. Sci., USA* **4** (1997): 132–138.

47. Mansky, K. C., A. Batiza, and P. F. Lambert. "Bovine Papillomavirus Type 1 E1 and Simian Virus 40 Large T Antigen Share Regions of Sequence Similarity Required for Multiple Functions." *J. Virol.* **71** (1997): 7600–7608.

48. Massimi, P., and L. Banks. "Repression of p53 Transcriptional Activity by the HPV E7 Proteins." *Virology* **227** (1997): 255–259.

49. Matthews, M. B. "Interactions Between Viruses and the Cellular Machinery for Protein Synthesis." In *Translational Control*, edited by J. W. B. Hershey, M. B. Matthews, and N. Sonenberg, 505–548. Cold Spring Harbor: Cold Spring Harbor Laboratory Press, 1996.

50. McGeoch, D. J. "Protein Sequence Comparisons Show That the 'Pseudoproteases' Encoded by Poxviruses and Certain Retroviruses Belong to the Deoxyuridine Triphosphate Family." *Nucleic Acids Res.* **18** (1990): 4105–4110.

51. Melnick, J. L., W. E. Rawls, and E. Adam. "Cervical Cancer." In *Viral Infections of Humans: Epidemiology and Control*, edited by A. S. Evans, 687–711, 3d ed. New York: Plenum, 1989.

52. Messerschmitt, A. S., N. Dunant, and K. Ballmer-Hofer. "DNA Tumor Viruses and Src Family Tyrosine Kinases, an Intimate Relationship." *Virology* **227** (1997): 271–280.

53. Meyers, C., and L. A. Laimins. "*In Vitro* Systems for the Study and Propagation of Human Papillomaviruses." In *Human Pathogenic Papillomaviruses*, edited by H. zur Hausen, 199–215. Berlin: Springer-Verlag, 1994.

54. Michaelian, I., M. Krieg, M. J. Gait, and J. Karn. "Interactions of INS (CRS) Elements and the Splicing Machinery Regulate the Production of Rev-Responsive mRNAs." *J. Mol. Biol.* **257** (1996): 246–264.

55. Moss, B. "Regulation of Vaccinia Virus Transcription." *Ann. Rev. Biochem.* **59** (1990): 661–688.

56. Myers, G. "HIV: Between Past and Future." *AIDS Research and Human Retroviruses* **10** (1994): 1317–1323.

57. Myers, G., and G. N. Pavlakis. "Evolutionary Potential of Complex Retroviruses." In *The Retroviridae*, edited by J. A. Levy, 51–105, vol. I. New York: Plenum, 1992.

58. Myers, G., H. Lu, C. Calef, and T. Leitner. "Heterogeneity of Papillomaviruses." *Sem. Cancer Biol.* **7** (1997): 349–358.

59. Myers, G., B. Korber, B. Foley, K.-T. Jeang, J. W. Mellors, and. S. Wain-Hobson, eds. *Human Retroviruses and AIDS 1996.* Los Alamos, NM: Theoretical Division, Los Alamos National Laboratory, 1996.

60. Narayan, O., and J. E. Clements. "Lentiviruses." In *Virology,* edited by B. N. Fields and D. M. Knipe, 1571–1585, 2nd ed. New York: Raven Press, 1990.

61. Nowak, M., and P. Schuster. "Error Threshold of Replication in Finite Populations." *J. Theor. Biol.* **137** (1989): 375–395.

62. O'Conner, M., S.-Y. Chan, and H.-U. Bernard. "Transcription Factor Binding Sites in the Long Control Region of Genital HPVs." In *Human Papillomaviruses 1995,* edited by G. Myers et al., III-21–III-40. Los Alamos, NM: Theoretical Division, Los Alamos National Laboratory, 1995.

63. Oldstone, M. B. A. "How Viruses Escape from Cytotoxic T Lymphocytes: Molecular Parameters and Players." *Virology* **234** (1997): 17–185.

64. Pavlakis, G. N. "The Molecular Biology of Human Immunodeficiency Virus Type 1." In *AIDS: Biology, Diagnosis, Treatment and Prevention,* edited by V. T. DeVita, Jr., S. Hellman, and S. A. Rosenberg, 45–74, 4th ed. New York: Lippencott-Raven, 1997.

65. Pennie, W. D., G. J. Grindlay, M. Cairney, and M. S. Campo. "Analysis of the Transforming Functions of Bovine Papillomavirus Type 4." *Virology* **193** (1993): 614–620.

66. Perelson, A. S., P. Essunger, and D. D. Ho. "Dynamics of HIV-1 and CD4+ Lymphocytes *In Vivo*." *AIDS* **11** (Suppl. A) (1997): S17–S24.

67. Pfister, H., and P. G. Fuchs. "Anatomy, Taxonomy, and Evolution of Papillomaviruses." *Intervirology* **37** (1994): 143–149.

68. Philipson, L. "Adenovirus—An Eternal Archetype." In *The Molecular Repertoire of Adenoviruses I,* edited by W. Doerfler and P. Bohm. 1–24. Berlin: Springer-Verlag, 1995.

69. Radetsky, P. *The Invisible Invaders.* Boston: Little, Brown and Co., 1994.

70. Re, F., D. Braaten, E. K. Franke, and J. Luban. "Human Immunodeficiency Virus Type 1 Vpr Arrests the Cell Cycle in G2 by Inhibiting the Activation of p34cdc2-cyclin b." *J. Virol.* **69** (1995): 6859–6864.

71. Reddy, T. R., G. Kraus, O. Yamada, D. J. Looney, M. Suhasini, and F. Wong-Staal. "Comparative Analyses of Human Immunodeficiency Virus Type 1 (HIV-1) and HIV-2 *vif* Mutants." *J. Virol.* **69** (1995): 3549–3553.

72. Refaeli, Y., D. N. Levy, and D. B. Weiner. "The Glucocorticoid Receptor Type II Complex Is a Target of the HIV-1 Vpr Gene Product." *Proc. Natl. Acad. Sci., USA* **92** (1995): 3621–3625.

73. Rittner, K., M. J. Churcher, M. J. Gait, and J. Karn. "The Human Immunodeficiency Virus Long Terminal Repeat Includes a Specialised Initiator Element Which Is Required for Tat-Responsive Transcription." *J. Mol. Biol.* **248** (1995): 562–580.

74. Rosenblatt, J. D., S. Miles, J. C. Gasson, and D. Prager. "Transactivation of Cellular Genes by Human Retroviruses." In *Transacting Functions of Human Retroviruses*, edited by I. S. Y. Chen, H. Koprowski, A. Srinivasan, and P. K. Vogt, 25–50. Berlin: Springer-Verlag, 1995.

75. Ruprecht, R., T. W. Baba, R. Rasmussen, Y. Hu, and P. L. Sharma. "Murine and Simian Retrovirus Models: The Threshold Hypothesis." *AIDS* **10** (Suppl. A) (1996): S33–S40.

76. Salmon, J., N. Ramoz, P. Cassonnet, G. Orth, and F. Breitburd. "A Cottontail Rabbit Papillomavirus Strain (CPRVb) with Strikingly Divergent E6 and E7 Oncoproteins: An Insight in the Evolution of Papillomaviruses." *Virology* **235** (1997): 228–234.

77. Scheffner, M., H. Romanczuk, K. Munger, J. M. Huibregtse, J. A. Mietz, and P. Howley. "Functions of Human Papillomavirus Proteins." In *Human Pathogenic Papillomaviruses*, edited by H. zur Hausen, 83–100. Berlin: Springer-Verlag, 1994.

78. Schneider, R., M. Campbell, G. Nasioulas, B. K. Felber, and G. N. Pavlakis. "Inactivation of the Human Immunodeficiency Virus Type 1 Inhibitory Elements Allows Rev-Independent Expression of Gag and Gag/Protease and Particle Formation." *J. Virol.* **71** (1997): 4892–4903.

79. Schwartz, S., B. K. Felber, and G. N. Pavlakis. "Expression of Human Immunodeficiency Virus Type 1 *vif* and *vpr* mRNAs Is Rev-Dependent and Regulated by Splicing." *Virology* **183** (1991): 677–686.

80. SenGupta , D. N., and R. H. Silverman. "Activation of Interferon-Regulated dsRNA-Dependent Enzymes by HIV-1 Leader RNA." *Nucleic Acids Res.* **17** (1989): 969–978.

81. SenGupta, D. N., B. Berkhout, A. Gatignol, A. Zhou, and R. H. Silverman. "Direct Evidence for Translation Regulation by Leader RNA and Tat Protein in Human Immunodeficiency Virus Type 1." *Proc. Natl. Acad. Sci., USA* **87** (1990): 7492–7496.

82. Shamanin, V., H. zur Hausen, D. Lavergne, C. M. Proby, I. M. Leigh, C. Neumann, H. Hamm, M. Goos, U.-F. Haustein, E. G. Jung, G. Plewig, H. Wolff, and E.-M. de Villiers. "Human Papillomavirus Infections in Nonmelanoma Skin Cancers from Renal Transplant Recipients and Nonimmunosuppressed Patients." *J. Natl. Cancer Inst.* **88** (1996): 802–811.

83. Stewart, S. A., B. Poon, J. B. Jowett, and I. S. Chen. "Human Immunodeficiency Virus Type 1 Vpr Induces Apoptosis Following Cell Cycle Arrest." *J. Virol.* **71** (1997): 5579–5592.

84. Strebel, K. "Structure and Function of HIV-1 Vpu." In *Human Retroviruses and AIDS 1996*, edited by G. Myers, B. Korber, B. Foley, K.-T. Jeang, J. W. Mellors, and S. Wain-Hobson, III-19–III-27. Los Alamos, NM: Theoretical Division, Los Alamos National Laboratory, 1996.

85. Stivahtis, G. L., M. A. Soares, M. A. Vodicks, B. H. Hahn, and M. Emerman. "Conservation and Host Specificity of Vpr-Mediated Cell Cycle Arrest Suggest a Fundamental Role in Primate Lentivirus Evolution and Biology." *J. Virol.* **71** (1997): 4331–4338.

86. Sverdrup, F. and G. Myers. "The E1 Proteins." In *Human Papillomaviruses 1997*, edited by G. Myers et al., III-37–III-53. Los Alamos, NM: Theoretical Division, Los Alamos National Laboratory, 1997.

87. Svitkin, Y. V., A. Pause, and N. Sonenberg. "La Autoantigen Transactivates Translation Initiation Mediated by the 5′ Leader Sequence of the Human Immunodeficiency Virus Type 1 mRNA." *J. Virol.* **68** (1994): 7001–7007.

88. Tan, W., and S. Schwartz. "The Rev Protein of Human Immunodeficiency Virus Type 1 Counteracts the Effect of an AU-rich Negative Element in the Human Papillomavirus Type 1 Late 3′ Untranslated Region." *J. Virol.* **69** (1995): 2932–2945.

89. Thomas, L. *The Lives of the Cell: Notes of a Biology Watcher*, 5. New York: Viking Press, 1974.

90. Tristem, M., C. Marshall, A. Karpas, J. Petrik, and F. Hill. "Origin of Vpx in Lentiviruses." *Nature* **347** (1990): 341–342.

91. Van Ranst, M., R. Tachezy, and R. D. Burk. "Human Papillomaviruses: A Neverending Story?" In *Papillomavirus Reviews: Current Research on Papillomaviruses*, 1–20. Leeds: Leeds University Press, 1996.

92. Wain-Hobson, S. "Is Antigenic Variation of HIV Important for AIDS and What Might Be Expected in the Future?" In *The Evolutionary Biology of Viruses*, edited by S. S. Morse, 185–220. New York: Raven Press, 1994.

93. Wang, C., and A. Siddiqui. "Structure and Function of the Hepatitis C Virus Internal Ribosome Entry Site." In *Cap-Independent Translation*, edited by P. Sarnow, 99–116. Berlin: Springer-Verlag, 1995.

94. Wang, L., S. Mukherjee, F. Jia, O. Narayan, and L. J. Zhao. "Interaction of Virion Protein Vpr of Human Immunodeficiency Virus Type 1 with Cellular Transcription Factor Sp1 and Transactivation of Viral Long Terminal Repeat." *J. Biol. Chem.* **270** (1995): 25564–25569.

95. Wills, C. *Yellow Fever, Black Goddess*. Reading, MA: Addison-Wesley, 1996.

96. Yang, L., G. F. Morris, J. M. Lockyer, M. Lu, Z. Wang, and C. B. Morris. "Distinct Transcriptional Pathways of TAR-Dependent and TAR-Independent Human Immunodeficiency Virus Type-1 Transactivation by Tat." *Virology* **235** (1997): 48–64.

97. Zambetti, G. P., and A. J. Levine. "A Comparison of the Biological Properties of Wild-Type and Mutant p53." *FASEB J.* **7** (1993): 855–865.

98. Zauli, G., D. Gibellini, D. Milani, M. Massoni, P. Borgatti, M. Laplaca, and S. Capitani. "Human Immunodeficiency Virus Type 1 Tat Protein Protects Lymphoid, Epithelial and Neuronal Cell Lines from Death by Apoptosis." *Cancer Res.* **53** (1993): 4481–4485.

99. Zinkernagel, R. M. "Virus-Induced Immunopathology. " In *Viral Pathogenesis*, edited by N. Nathanson et al., 163–179. Philadelphia: Lippencott-Raven, 1997.

Alison A. McBride,* Elliot J. Androphy,† and Karl Münger‡
*Laboratory of Viral Diseases, National Institute of Allergy and Infectious Diseases, National
Institutes of Health, Bethesda, MD 20892–0455
†Department of Dermatology, New England Medical Center, and Department of Molecular
Biology and Microbiology, Tufts University, School of Medicine, Boston, MA 02111
‡Department of Pathology, Harvard Medical School, Boston, MA 02115–5701

Regulation of the Papillomavirus E6 and E7 Oncoproteins by the Viral E1 and E2 Proteins

The papillomaviruses infect and replicate in the stratified layers of skin and mucosa (epithelium), giving rise to benign lesions called warts or papillomas. The virus infects basal (the lowermost) epithelial cells, and these cells maintain low levels of viral DNA. Viral DNA amplification and virion assembly only occur as the cells of a papilloma differentiate in the upper layers of a stratified epithelium. In addition to the cell's replication enzymes, the E1 and E2 viral proteins are necessary for initiation of viral DNA replication. However, the differentiated cells in which viral DNA amplification begins are not dividing and so do not normally express the factors involved in DNA synthesis. The viral E6 and E7 proteins overcome this obstacle by inducing cells to enter S-phase and by abrogating cell-cycle controls. In addition to their roles in DNA replication, the E1 and E2 proteins regulate viral gene expression, including the promoter that transcribes the E6 and E7 genes. This chapter will briefly describe the functions of these viral proteins and provide an integrated view of current theories concerning their role in viral replication and malignant progression.

Viral Regulatory Structures and Their Degeneracies, edited by Gerald Myers.
SFI Studies in the Sciences of Complexity, Proc. Vol. XXVIII, Addison-Wesley, 1998

PAPILLOMAVIRUSES—REPLICATION OR CARCINOGENESIS?

The majority of papillomaviruses induce self-limiting, benign, proliferative epithelial lesions (Figure 1). A specific subset of papillomaviruses that infect the human anogenital tract (high-risk viruses) are associated with cancer, particularly that of

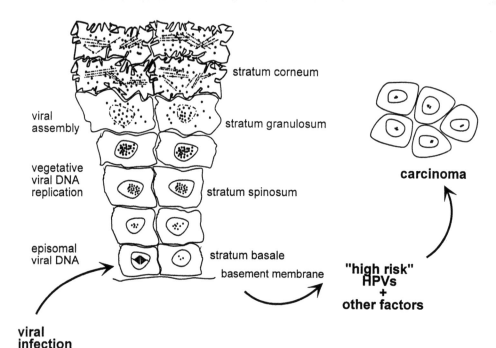

FIGURE 1   Diagram of differentiating cells in a stratified epithelium, and expression of viral functions in a papilloma. Basal cells form a single layer on the basement membrane and are the only cells that proliferate in normal skin. As these cells migrate upward, they differentiate and synthesize increased amounts of high molecular-weight keratins. The spinous layer of cells is characterized by intracellular keratin filaments that link to a protein complex which forms attachments between the cells. The attachments give the cells a spiny appearance and enable the skin to resist abrasion. The granular layer contains granules that fuse with the cell membrane and deposit material between cells to form an impenetrable barrier. Cells of the stratum corneum have no nuclei and contain mostly keratin bundles. Papillomaviruses infect basal skin cells in which they express early gene products. (Early viral gene expression is defined as that which takes place before viral DNA replication; late gene expression occurs after DNA replication.) In a papilloma, viral DNA amplification and late gene expression only occur in differentiating cells. In a small percentage of papillomas infected with high-risk viruses, cells can progress to form a carcinoma that produces no virus particles.

the uterine cervix (reviewed in zur Hausen[28]). Only a small percentage of lesions infected with high-risk viruses will progress to malignancy, implying that papillomavirus infection alone is not sufficient for malignant progression but predisposes the cells to accumulation of cellular mutations. In these cases, there is a gradual progression over a time frame of years to decades during which the papilloma may manifest disorganized differentiation (premalignant or dysplastic state) and may finally develop into an invasive cancer. From these observations it has been concluded that specific genes of the high-risk papillomaviruses and other cofactors are necessary for malignant progression. HPV-associated carcinomas no longer produce viral particles but invariably express the viral E6 and E7 proteins. In general, viral DNA is found to be integrated into cellular chromosomes, although no specific locus has been identified in the host genome. In the majority of cervical cancers, the viral genome is integrated such that the E1 and/or E2 open reading frames are disrupted (see Figure 2). This has led to the hypothesis that disruption of the regulatory functions of the E1 and E2 proteins (together with continued expression of E6 and E7) is a critical step in progression to a carcinoma.

THE VIRAL LIFE CYCLE

Papillomaviruses infect the basal layer of cells of a stratified epithelium (Figure 1). The a6b4 integrin protein, expressed exclusively in this cell layer, probably acts as a receptor for the virus.[6] Damage to the superficial skin or mucosa is presumed necessary to allow access of virus to the basal layer. Within basal cells, the viral genome is thought to be maintained as an extrachromosomal closed circle of double-stranded DNA (an episome, Figure 2). The viral episome is probably amplified to a low copy number in basal cells, and this is thought to require the viral E1 and E2 replication proteins. Presumably these proteins are expressed early after infection, since there is no evidence that they are carried within the viral particle. The viral E5 protein is also expressed in basal cells (Figure 3(a)). This protein stimulates activity of growth-factor receptors expressed by the cell and may provide signals to induce cell proliferation (reviewed in Howley[10]). Enhanced replication of basal cells may be important to increase the population of infected cells and to provide a suitable environment for establishment of a productive viral lesion. As basal cells differentiate and migrate up to the stratum spinosum, expression of the E2 proteins is greatly increased and vegetative DNA replication begins (Figure 3(a)–(b), reviewed in Howley[10] and McBride and Myers[18]). The E7 protein is also required at this stage to induce cells to enter S-phase and synthesize cellular replication proteins. However, this disruption of cell-cycle control in differentiating cells can be recognized by the host cell; such conflicting signals may signal to a cell to undergo apoptosis (programmed cell death to eliminate damaged cells) and one function of the E6 protein may be to prevent this (reviewed in Kubbutat and Vousden[16]). The E4 protein is abundant in the more differentiated layers of a papilloma. It has been hypothesized that E4 may function as a nuclear structural protein, an RNA splicing and transport factor, or in release of viral particles from the papilloma (reviewed in Howley[10]). In the upper-differentiated layers of the papilloma, the viral capsid proteins L1 and L2 are synthesized (Figure 3(c)) and virions are assembled (reviewed in Howley[10]).

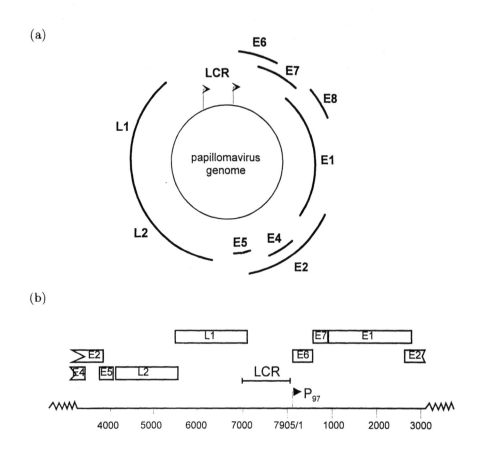

FIGURE 2   (a) Circular genomic map of a papillomavirus. The early open reading frames (E1–E8) and late open reading frames (L1 and L2 are indicated). The LCR (long control region) contains regulatory elements for transcription and DNA replication. (b) Map of a HPV-16 genome integrated into cellular chromosomes as is often found in cervical carcinomas. The genome has integrated in the E2 gene. Cellular DNA is represented by a jagged line.

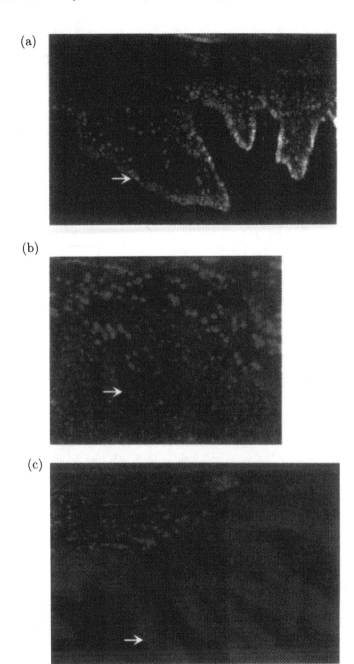

FIGURE 3 Localization of viral gene products in a fibropapilloma caused by BPV1 infection. In each case the basal layer of cells is indicated by an arrow. (a) E5 protein is stained red and the E2 proteins are stained green by immunofluorescence. (b) BPV1 DNA is stained green by fluorescent *in situ* hybridization. (c) Viral structural proteins (L1 and L2) are stained green by immunofluorescence.

TRANSCRIPTIONAL REGULATION BY THE VIRAL E2 PROTEINS

In bovine papillomavirus type 1 several gene products are expressed from the E2 open reading frame (ORF) and they have been shown to function as transcriptional activators and repressors (Figure 4, reviewed in Fuchs and Pfister,[9] Howley,[10]

(a) **BPV1**

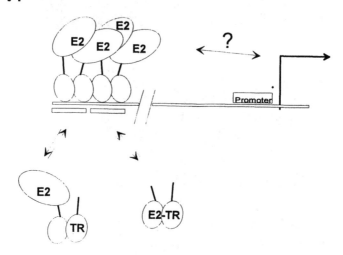

(b) **HPV16**

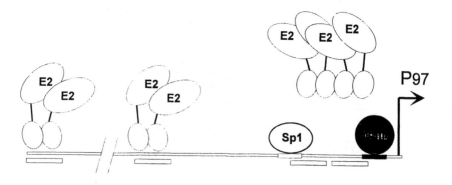

FIGURE 4 Putative mechanisms of transcriptional regulation by the papillomavirus E2 proteins. (a) BPV1 expresses a transcriptional transactivator with a transactivation domain and DNA binding/dimerization domain. (continued)

FIGURE 4 (continued) Two shorter repressor proteins contain only the DNA binding/dimerization domain and are thought to repress E2 transactivation by forming heterodimers with the *trans*-activator and by competing for the binding to the E2 sites in the viral genome. (b) It has been proposed that in some HPVs the full-length E2 protein can activate transcription by interacting with higher-affinity E2 binding sites upstream from the transcriptional start site. At higher levels of E2, the lower-affinity sites close to the promoter become occupied. This displaces essential cellular factors, SP1 and TFIId, and results in repression of basal promoter activity.

and McBride and Myers[18]). cDNA species, copies of messenger RNA, that could potentially encode truncated human papillomavirus E2 repressor proteins have been cloned but, as yet, no such proteins have been identified. Some HPVs may have evolved a mechanism to both activate and repress viral transcription with the full-length E2 protein (see Figure 4, reviewed in Fuchs and Pfister[9] and McBride and Myers[18]). The full-length E2 protein from all papillomaviruses consists of a 200-amino-acid *N*-terminal transactivation domain linked to a 100-amino-acid *C*-terminal DNA binding and dimerization domain by a region of variable length and sequence. The E2 proteins activate transcription by binding to specific E2 DNA binding sites located near viral promoters. The exact mechanism of transactivation has not been elucidated but probably involves interaction with components of the basic transcriptional machinery. BPV1 E2 has been shown to interact with SP1, TBP, TFIIB, and a novel cellular protein, AMF-1[2] (reviewed in McBride and Myers[18]). In several HPVs associated with the anogenital tract, the full-length E2 protein appears to repress the promoter located upstream from the E6 gene (reviewed in Fuchs and Pfister[9] and McBride and Myers[18]). This probably occurs when the E2 dimer binds to E2 DNA binding sites, which overlap binding sites for the cellular SP1 and TFIID transcription factors. Recent studies have indicated that these proximal E2 binding sites have lower affinity for the E2 protein than the E2 binding sites located further upstream from the promoter start site.[25] This has led to a model in which low levels of E2 bind to the higher-affinity upstream E2 sites and activate transcription, but at high levels of E2 protein the lower-affinity proximal E2 sites are occupied leading to transcriptional repression. It is likely that the situation is more complex in a papilloma, as levels and activities of E2 proteins and cellular transcription factors are probably modulated by cell cycle and epithelial differentiation.

THE HPV E6 AND E7 PROTEINS CAN REGULATE CELL CYCLE

It was initially proposed that HPV E6 and E7 were oncogenes (cancer causing genes) following the observation that these genes were selectively retained and expressed in

HPV positive cervical cancers and that their continued expression was required for proliferation of carcinoma-derived lines (reviewed in zur Hausen[28]). Normal human cells stop growing after a finite time in culture but expression of the E6 and E7 proteins allows them to continue to divide. This immortalization is thought to be one step in the progression to full malignancy. The E6 and E7 proteins of the high-risk HPVs cooperate to immortalize primary keratinocytes (epithelial cells). They do this by interacting with tumor suppressor proteins that regulate the cell cycle, such as p53 and the retinoblastoma gene product, Rb. The DNA tumor viruses SV40 and adenovirus also disrupt cell-cycle regulation in a similar fashion; the large tumor antigen (TAg) of SV40 and the E1B and E1A gene products of adenovirus also interact with and inactivate the functions of p53 and Rb. The interaction of E6 and E7 with these important targets is thought to be responsible for their biological effects, although this has yet to be proven experimentally.

THE HPV E7 PROTEINS INTERACT WITH NEGATIVE REGULATORS OF CELL GROWTH

The HPV E7 proteins share functional and structural similarities with oncoproteins encoded by other small DNA tumor viruses, including TAgs of the polyomaviruses and the E1A proteins of the adenoviruses. The high-risk HPV E7 proteins function as oncogenes in multiple assays and together with E6 can immortalize primary human fibroblasts and keratinocytes (reviewed in Kubbutat and Vousden[16]). The sequences that are conserved among the viral E7 oncoproteins are critical for their growth-transforming activities and participate in complex formation with a conserved set of negative-growth regulatory proteins, including Rb. These cellular proteins are involved in regulating the activities of the E2F family of proteins, which function as transcriptional activators and repressors of many genes that encode enzymes that contribute to S-phase progression and DNA replication (see Figure 5). In the G1 phases of the cell-division cycle, Rb is present in a hypophosphorylated form and can interact with E2F. This Rb/E2F complex functions as a transcriptional repressor and dissociates when Rb is phosphorylated by cyclin-dependent kinase complexes near the G1 to S boundary. Free E2F acts as a transcriptional activator, and the regulated conversion of E2F between transcriptional repressor and activator is necessary for regulated G1/S-phase transition. The binding of HPV E7 proteins to Rb and related proteins disrupts Rb/E2Fcomplexes and leads to synthesis of E2F-responsive S-phase genes. Recent studies have shown that E7 can not only interact with and inactivate Rb but can also induce its degradation.[1,15] This disruption of the delicate, cell-cycle-dependent regulation of the transcriptional activity of E2F is an important way in which E7 contributes to the reactivation of DNA synthesis in normally nonreplicating cells of the epithelium.

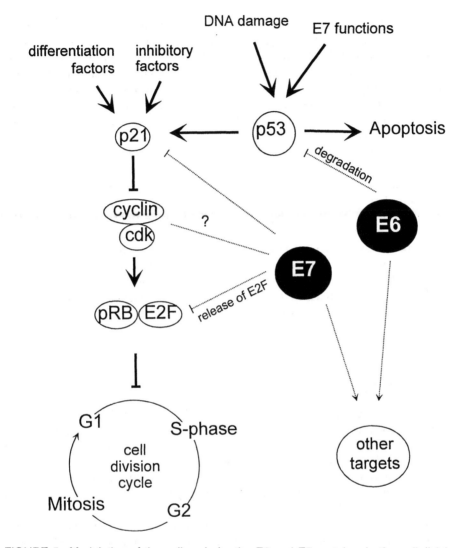

FIGURE 5 Modulation of the cell cycle by the E6 and E7 proteins. In the cell-division cycle, chromosomes are duplicated by DNA synthesis in S-phase and segregated to two daughter cells in mitosis. These events are separated by two periods of growth and reorganization (G1 and G2). Progress through the cell cycle is regulated by many cyclin-cdk (cyclin-dependent kinase) complexes. E7 promotes cellular DNA synthesis by interfering at several steps of the pathway that regulates the transition from G1 to S-phase. The cellular p53 protein is one of the key regulators of the checkpoint between G1 and S-phase, and the unscheduled E7-mediated cellular proliferation induces activity of p53 resulting in cell-cycle arrest or apoptosis (programmed cell death). The E6 protein abrogates this response by degrading p53 and thus allowing cells to proceed into S-phase.

The HPV E7 protein can also bind to a cellular protein kinase activity that is thought to contain, at least in part, cyclin E and/or cyclin A and cdk2 (reviewed in Kubbutat and Vousden[16]). It is possible that by interacting with such kinase complexes, E7 may influence their substrate specificities or interfere with their regulation by effector molecules such as cdk-inhibitors. Indeed, it has been shown that the HPV E7 protein can interact with cdk-inhibitors. Similar to the adenovirus E1A protein, HPV E7 can reverse inhibition of cdk2 activity by p27Kip1[27] and p21cip1.[14] Thus, HPV E7 is able to maintain high levels of cdk activity and promote cell-cycle progression in the presence of negative-growth regulatory stimuli. This may be the mechanism by which HPV-16 E7 promotes DNA synthesis in the presence of antiproliferative signals, the lack of growth factors, loss of cell adhesion, or induction of keratinocyte differentiation.

In a productive papilloma, expression of E7 may be tightly regulated so that there is a fine balance between cellular proliferation and differentiation. This balance is important so that viral DNA replication can occur, yet allow these cells to continue to differentiate in order to express late genes and assemble progeny virus. Changes in the control of E7 gene expression (for example, by integration of viral DNA) may disrupt this balance, leading to further deregulation of cell-cycle control and ultimately malignant progression.

THE E6 PROTEIN

The majority of E6 studies involve analyses of the high-risk HPV types that are commonly associated with cervical cancer. The activities of high-risk HPV E6 are most evident in primary cell immortalization. Together with HPV E7, the E6 protein alters the differentiation program of epithelial cells in organotypic culture (a culture system that supports differentiation of a stratified epithelium[19]) and efficiently immortalizes primary human keratinocytes (reviewed in Kubbutat and Vousden[16]). Some studies show that HPV-16 E6 renders keratinocytes resistant to calcium- and serum-induced terminal differentiation. This is presumed to reflect the need of the virus to interfere with completion of the normal epithelial terminal differentiation program, which might block the cell's ability to synthesize the factors necessary for DNA replication.

The E6 and E7 proteins act in concert to regulate the cell cycle. Agents that damage DNA or deregulate cell cycle, including HPV E7, cause an increase in p53 levels and can predispose cells to undergo apoptosis. p53 is a transcriptional activator of genes involved in G1 arrest, apoptosis, and DNA repair. An increase in p53 levels can lead to growth arrest or cell death, and one of the primary functions of E6 may be to prevent this. HPV E6 can inactivate p53 and inhibit its transcriptional activity (reviewed in Kubbutat and Vousden[16]). High-risk HPV E6 complexes with p53 but the ability of the low-risk E6 proteins to bind p53 is controversial. Incubation of high-risk E6 with p53 leads to its selective degradation, which reduces the half life of p53 and disrupts p53-mediated G1 arrest. The degradation reaction is ATP-dependent, and p53 becomes covalently linked to ubiquitin. A cellular factor called E6-AP (E6-associated protein) is necessary for HPV-16 E6 binding to and degradation of p53. The E6/E6-AP complex acts as an ubiquitin ligase (reviewed in Howley[10]).

The high-risk HPV-16 E6 protein also interacts with other cellular factors, including one which has been designated E6BP. This factor is identical to ERC-55, a calcium-binding protein that resides in the endoplasmic reticulum. Calcium ions and calcium-binding proteins are important for epithelial cellular proliferation and differentiation. One hypothesis is that E6 might inhibit cellular differentiation through its interaction with the E6BP calcium-binding protein.[3] E6 also binds paxillin, a protein involved in coupling signal transduction to the cytoskeleton.[26] The biological significance of these interactions remains to be determined.

HPV E6 can stimulate telomerase activity in human keratinocytes (reviewed in Kubbutat and Vousden[16]), and this seems to be independent of p53 binding and/or degradation. The telomerase complex synthesizes the telomere repeat sequences at the ends of chromosomes, and its level often correlates with cellular immortalization. Nevertheless, the relationship between telomerase activity, telomere lengthening, and immortalization requires further investigation.

Most mucosal and cutaneous HPV (and BPV1) E6 proteins possess some cellular immortalizing or growth-transforming functions that are independent of the ability to bind and degrade p53 or inhibit p53-dependent activation of transcription (reviewed in Kubbutat and Vousden[16]). The E6 and E7 proteins of all papillomaviruses must provide functions necessary to ensure viral DNA synthesis in cells that are undergoing terminal differentiation. Therefore, the unique ability of the high-risk HPV E6 proteins to degrade p53 must not be associated with their ability to interfere with differentiation. In each papillomavirus the E6 and E7 proteins have evolved together and may balance each other's activities. High-risk HPV E7 proteins bind Rb more efficiently than other HPV E7 proteins, and this may be more likely to induce cells to undergo apoptosis. The high-risk E6 proteins are thought to have evolved a mechanism to destroy p53 in order to protect against cell death and balance the effects of the high-risk E7 proteins. The p53 protein normally induces a checkpoint to allow the repair of genotoxic damage that results from radiation and chemicals. Cells can, therefore, accumulate genetic damage induced by these factors. A subset of the resulting cellular mutations may eventually lead to malignant progression.

VIRAL DNA REPLICATION AND THE E1 AND E2 PROTEINS

The E6 and E7 proteins are believed to establish conditions in which viral DNA synthesis can occur in the differentiating epithelial cells. The viral E1 and E2 proteins recruit the requisite replication complexes to the viral origin of replication and amplify the viral genomes (reviewed in Howley[10]).

Three stages of DNA replication take place in the papillomavirus life cycle. After the initial uptake of the virus, the particle is uncoated and the genome is transported to the nucleus of the basal cell where it is presumed to be amplified to a low copy number. Most experimental studies have studied transient DNA replication in cultured cells, which is probably most analogous to this initial amplification and which requires the E1 and E2 proteins and the viral replication origin. Presumably, a low level of the E1 and E2 proteins must be expressed early after infection since there is no evidence that they are in the viral particle. The viral replication origin consists of an E1 binding site, an E2 binding site, and an AT-nucleotide-rich region. The E1 and E2 sites have relatively low affinity for their respective proteins, but together E1 and E2 cooperatively bind to the origin with high affinity. After the initial binding of an E1/E2 complex to the origin, the E1 protein oligomerizes to form a trimer or hexamer that encircles the DNA, and E2 may dissociate from the origin.[7,23] The E1 protein functions as a helicase that unwinds the DNA at the origin to allow DNA synthesis to begin (Figure 6).

Stable episomal maintenance is the second stage of papillomavirus DNA replication. It requires the E1 and E2 proteins and, in addition to the origin of replication, regions of DNA from the LCR that contain several E2 binding sites suggesting that the E2 protein might play an important role in episomal maintenance.[20] Recent studies have shown that both the BPV1 E2 *trans*-activator protein and BPV genomes are associated with cellular chromosomes at mitosis.[24] This could be a mechanism by which approximately equal numbers of viral genomes are segregated to daughter cells at cell division to ensure that all basal cells of a papilloma contain viral DNA.

The third stage of viral replication is vegetative DNA replication which is required to generate progeny virus. As basal cells differentiate and migrate to the stratum spinosum of a papilloma, greatly increased amounts of viral DNA are synthesized as vegetative DNA replication begins (reviewed in Howley[10]). Increased expression of the E2 proteins also occurs in the stratum spinosum and may be important for amplification of viral DNA (see Figure 3(b)). The E2 protein is important for initiation of viral DNA replication, but it has also been shown that HPV-31 E2 can arrest cells in *S*-phase.[8] Clearly this could be important for vegetative replication because it allows sustained synthesis of viral DNA. Very little else is known about vegetative viral DNA replication because of the requirement for terminally differentiating keratinocytes. However, great advances are being made in replicating papillomaviruses in organotypic raft cultures and in xenografts of mice, and these systems should prove to be very useful in studying the entire viral life cycle (reviewed in Meyers and Laimins[19]).

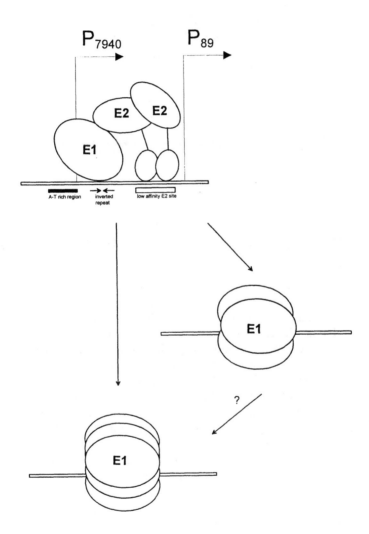

FIGURE 6 Model of initiation of viral DNA replication. The E1 and E2 proteins initiate DNA replication by cooperatively binding to specific sites in the viral origin of replication. It has been proposed that an E1-alone complex then assembles in a ring-like structure around the DNA (this may be a trimer or a hexamer) and the helicase activity of E1 unwinds the origin to allow access of the cellular replication machinery.

There are some indications that the viral E1 replication protein may also function as a transcriptional repressor. Inactivation of E1 can increase the immortalizing or growth transforming potential of HPV-16 and BPV1, respectively (reviewed in Scheffner et al.[22]), and this correlates well with the frequent disruption of E1 and/or E2 expression found in HPV-associated carcinomas. The E1 protein of BPV1 can repress E2-mediated transactivation of the viral P_{89} promoter, which expresses the E6 and E7 gene products.[17,22] As shown in Figure 6, this is probably a consequence of binding of an E1/E2 complex to the replication origin that is located just upstream from P_{89}. Perhaps the location of the origin in the vicinity of one of the major viral promoters serves to make the processes of transcription and replication mutually exclusive. In the presence of the E1 protein, an E1/E2 complex could bind to the origin and initiate replication but transcription would be inhibited. Conversely, in the absence of E1, the E2 protein would interact with other E2 sites and regulate transcription.

CONSEQUENCES OF DISRUPTING E1 AND/OR E2 GENE EXPRESSION

HPV-associated genital carcinomas constitutively express the E6 and E7 proteins and require these proteins for continued proliferation and maintenance of the transformed state. (Malignant transformation is a change in cells in culture that is characterized by changes in growth properties and morphology and usually increases their ability to cause tumors in animals.) The viral genome is often integrated in these carcinomas in such a way as to disrupt expression of the E1 and/or E2 proteins (Figure 2, reviewed in Howley[10]). It is generally believed that disruption of the regulatory properties of the E1 and E2 proteins results in deregulated expression of the E6 and E7 proteins. In a productive lesion there is likely a fine balance in the levels of E6 and E7 required to induce S-phase in differentiating cells so that viral genomes can be amplified, yet still allow these cells to continue to differentiate in order that late viral functions can be expressed. Deregulation of E6 and E7 gene expression may disrupt this equilibrium in favor of cellular proliferation; cells constitutively expressing E6 and E7 proteins can proliferate bypassing many of the normal cell-cycle checkpoints and so can accumulate genetic damage that could ultimately result in malignant progression.

There is circumstantial evidence indicating that disrupted expression of the E1 and/or E2 proteins is important for deregulation of E6 and E7 gene expression and for malignant progression but, as yet, no single model has been defined to explain these observations. HeLa cells, which are derived from an HPV-18-containing cervical carcinoma, constitutively express the viral E6 and E7 proteins. E6 and E7 gene expression is required for cellular proliferation, but can be repressed by introduction of an E2 gene that causes the cells to undergo growth arrest and/or apoptosis

(Desaintes et al.[4] and references therein). The W12 cell line, which is also derived from a cervical lesion, contains subpopulations of cells containing either episomal HPV-16 or HPV-16 integrated in the E2 gene; the latter population has a growth advantage over those cells containing only episomal viral DNA.[12] The capacity of HPV-16 to induce differentiation-resistant colonies of primary keratinocytes can also be greatly increased using genomes containing translational stops within the E1 or E2 genes.[21]

Two models have been proposed to explain the effect of E2 inactivation on E6 and E7 gene expression. The E2 *trans*-activator protein can repress the promoter located upstream from the E6 gene in genital mucosal-associated HPVs, and so inactivation of E2 is predicted to increase transcription from this promoter. A similar phenomenon might occur with inactivation of E1, as this has also been shown to be able to repress viral transcription. However, integration within E1 would also eliminate expression from the downstream E2 gene. Notably, in the high-risk viruses, both E6 and E7 proteins are encoded by differentially spliced transcripts synthesized from the promoter located just upstream from E6 at the 3' end of the LCR. Derepression of this promoter should result in deregulated expression of both viral proteins. In contrast, the low-risk viruses, such as HPV-6 and HPV-11, are thought to express E6 from an E2-regulated promoter analogous to that of HPV-16 and HPV-18, and E7 from a different promoter that initiates within the E7 gene (reviewed in Fuchs and Pfister[9]). In this case, disruption of E2 regulation could potentially deregulate the expression of E6 but not that of the E7 protein. The E6 and E7 proteins are expressed in all cells of a carcinoma, but there is some controversy as to which layer of the epithelium they are expressed in a productive lesion. Some studies find the proteins to be expressed only in suprabasal cells, but others find them throughout the lesion. In the former case, an additional mechanism of E6 and E7 deregulation could involve the inappropriate expression of E6 and E7 in the basal cells, which may be more susceptible to cell cycle deregulation and genomic instability.

It has also been proposed that E6 and E7 gene expression can be regulated by mRNA stability. An element has been identified in the 3' untranslated region of HPV-16 mRNAs that confers instability. Integration of the viral genome into the host chromosome would remove this element, resulting in increased stability of E6 and E7 mRNAs and their encoded proteins.[13]

There may be several different ways in which E6 and E7 gene expression can be deregulated; it is unlikely to be due to one simple process, such as E2-mediated transcriptional repression, as carcinomas can also contain uninterrupted E2 (often episomal). For example, HPV regulatory regions contain negative regulatory elements for the cellular transcription factor YY1, and often some of these are found to be mutated in carcinomas containing purely episomal DNA, resulting in up-regulation of the E6 promoter (reviewed in Fuchs and Pfister[9]). Clearly, further studies are required to understand the mechanisms by which E6 and E7 gene expression is regulated and how their deregulation can lead to genomic instability and carcinogenesis.

There is also some evidence that the E2 protein may have some growth inhibitory properties that are independent of E2-mediated regulation of the E6 and E7 promoter. E2 can induce growth arrest and apoptosis in HeLa cells (a cervical carcinoma-derived cell line containing HPV-18),[4,5] but some studies also find that E2 can inhibit growth of HPV-negative carcinoma lines.[11] In addition, genetic studies examining the effect of mutations of either E1 or E2 on the immortalizing capacity of the HPV-16 genome implicate additional mechanisms in the E2 repression of viral immortalization functions.[21] Therefore, inactivation of the E2 gene may release cell growth inhibition by E2 and may be intrinsically important to malignant progression.

ACKNOWLEDGMENTS

The authors apologize to those whose work or original publications could not be cited because of space limitations.

REFERENCES

1. Boyer, S. N., D. E. Wazer, and V. Band. "E7 Protein of Human Papillomavirus-16 Induces Degradation of Retinoblastoma Protein Through the Ubiquitin-Proteasome Pathway." *Cancer Res.* **56** (1996): 4620–4624.
2. Breiding, D., F. Sverdrup, M. J. Grossel, N. Moscufo, W. Boonchai, and E. J. Androphy. "Isolation of a BPV1 E2 Transactivation Domain Binding Factor Required for Both Transcriptional Activation and DNA Replication." (1997): Manuscript submitted.
3. Chen, J. J., C. E. Reid, V. Band, and E. J. Androphy. "Interaction of Papillomavirus E6 Oncoproteins with a Putative Calcium-Binding Protein." *Science* **269** (1995): 529–531.
4. Desaintes, C., C. Demeret, S. Goyat, M. Yaniv, and F. Thierry. "Expression of the Papillomavirus E2 Protein in HeLa Cells Leads to Apoptosis." *EMBO J.* **16** (1997): 504–514.
5. Dowhanick, J. J., A. A. McBride, and P. M. Howley. "Suppression of Cellular Proliferation by the Papillomavirus E2 Protein." *J. Virol.* **69** (1995): 7791–7799.
6. Evander, M., I. H. Frazer, E. Payne, Y. M. Qi, K. Hengst, and N. A. McMillan. "Identification of the Alpha6 Integrin as a Candidate Receptor for Papillomaviruses." *J. Virol.* **71** (1997): 2449–2456.
7. Fouts, E., E. Egelman, and M. Botchan. "E1 Protein of BPV1 Forms Hexameric Rings with a 30-40A Hole." (1997): Manuscript in preparation.
8. Frattini, M. G., S. D. Hurst, H. B. Lim, S. Swaminathan, and L. A. Laimins. "Abrogation of a Mitotic Checkpoint by E2 Proteins from Oncogenic Human Papillomaviruses Correlates with Increased Turnover of the p53 Tumor Suppressor Protein." *EMBO J.* **16** (1997): 318–331.
9. Fuchs, P. G., and H. Pfister. "Transcription of Papillomavirus Genomes." *Intervirology* **37** (1994): 159–167.
10. Howley, P. M. "Papillomavirinae: The Viruses and Their Replication." In *Virology*, edited by B. N. Fields, D. M. Knipe, and P. M. Howley, 2045–2076. Philadelphia and New York: Lippincott-Raven, 1995.
11. Hwang, E. S., D. J. Riese, II, J. Settleman, L. A. Nilson, J. Honig, S. Flynn, and D. DiMaio. "Inhibition of Cervical Carcinoma Cell Line Proliferation by the Introduction of a Bovine Papillomavirus Regulatory Gene." *J. Virol.* **67** (1993): 3720–3729.
12. Jeon, S., B. L. Allen-Hoffmann, and P. F. Lambert. "Integration of Human Papillomavirus Type 16 Into the Human Genome Correlates with a Selective Growth Advantage of Cells." *J. Virol.* **69** (1995): 2989–2997.
13. Jeon, S., and P. F. Lambert. "Integration of Human Papillomavirus Type 16 DNA into the Human Genome Leads to Increased Stability of E6 and E7

mRNAs: Implications for Cervical Carcinogenesis." *Proc. Natl. Acad. Sci. USA* **92** (1995): 1654–1658.

14. Jones, D. L., R. M. Alani, and K. Münger. "HPV-16 E7 can Reverse Inhibition of cdk2 Activity by p21cip1." (1997): Manuscript in preparation.

15. Jones, D. L., and K. Münger. "Analysis of the p53-mediated G1 Growth Arrest Pathway in Cells Expressing the Human Papillomavirus Type 16 E7 Oncoprotein." *J. Virol.* **71** (1997): 2905–2912.

16. Kubbutat, M. H. G., and K. H. Vousden. "Role of E6 and E7 Oncoproteins in HPV-Induced Anogenital Malignancies." *Seminars in Virology* **7** (1996): 295–304.

17. Le Moal, M. A., M. Yaniv, and F. Thierry. "The Bovine Papillomavirus Type 1 (BPV1) Replication Protein E1 Modulates Transcriptional Activation by Interacting with BPV1 E2." *J. Virol.* **68** (1994): 1085–1093.

18. McBride, A. A., and G. Myers. "The E2 Proteins." In *Human Papillomaviruses*, edited by G. Myers, C. Baker, C. Wheeler, A. Halpern, A. McBride, and J. Doorbar. Los Alamos, NM: Theoretical Division, Los Alamos National Laboratory, 1996.

19. Meyers, C., and L. A. Laimins. "*In Vitro* Systems for the Study and Propagation of Human Papillomaviruses." *Curr. Top. Microbiol. Immunol.* **186** (1994): 199–215.

20. Piirsoo, M., E. Ustav, T. Mandel, A. Stenlund, and M. Ustav. "*Cis* and Trans Requirements for Stable Episomal Maintenance of the BPV-1 Replicator." *EMBO J.* **15** (1996): 1–11.

21. Romanczuk, H., and P. M. Howley. "Disruption of Either the E1 or the E2 Regulatory Gene of Human Papillomavirus Type 16 Increases Viral Immortalization Capacity." *Proc. Natl. Acad. Sci. USA* **89** (1995): 3159–3163.

22. Scheffner, M., H. Romanczuk, K. Münger, J. M. Huibregtse, J. A. Mietz, and P. M. Howley. "Functions of Human Papillomavirus Proteins." *Curr. Top. Microbiol. Immunol.* **186** (1994): 83–99.

23. Sedman, J., and A. Stenlund. "The Initiator Protein E1 Binds to the Bovine Papillomavirus Origin of Replication as a Trimeric Ring-Like Structure." *EMBO J.* **15** (1996): 5085–5092.

24. Skiadopoulos, M. H., and A. A. McBride. "BPV1 Viral Genomes and the E2 Transactivator Protein are Associated with Cellular Metaphase Chromosomes." (1997): Manuscript in preparation.

25. Steger, G., and S. Corbach. "Dose-Dependent Regulation of the Early Promoter of Human Papillomavirus Type 18 by the Viral E2 Protein." *J. Virol.* **71** (1997): 50–58.

26. Tong, X., and P. M. Howley. "The Bovine Papillomavirus E6 Oncoprotein Interacts with Paxillin and Disrupts the Actin Cytoskeleton." *Proc. Natl. Acad. Sci. USA* **94** (1997): 4412–4417.

27. Zerfass-Thome, K., W. Zwerschke, B. Mannhardt, R. Tindle, J. W. Botz, and P. Jansen-Durr. "Inactivation of the cdk Inhibitor p27KIP1 by the Human Papillomavirus Type 16 E7 Oncoprotein." *Oncogene* **13** (1996): 2323–2330.

28. zur Hausen, H. "Papillomavirus Infections—A Major Cause of Human Cancers." *Biochim. Biophys. Acta* **1288** (1996): F55–F78.

Françoise Thierry* and Moshe Yaniv
Unité des Virus Oncogènes, U1644 CNRS, Institut Pasteur, 28 rue du Dr Roux, 75724, Paris, Cedex 15, FRANCE; *E-mail: fthierry@pasteur.fr

The Control of Human Papillomavirus Transcription

1. INTRODUCTION

Papillomaviruses are a large family of viruses which contain double-stranded DNA and infect higher vertebrates. This class of viruses exclusively causes lesions of the epithelium of the skin and mucous. We will focus on a subclass of HPVs (human papillomaviruses) that have been found associated with lesions of the anogenital tract. These so-called "genital papillomaviruses" are widespread in nature where they are venereally transmitted. While in most cases, lesions of the cervix caused by these viruses remain benign and regress, in some individuals they progress to malignant tumors. Squamous cell carcinoma of the cervix is the second most common cancer in women after breast cancer worldwide and the first in underdeveloped countries. In more than 80% of cases it is associated with HPV infection and viral DNA persists in the malignant cells. Lesions containing certain subtypes of genital papillomaviruses, such as type 16, 18, or 33, are more prone to malignant conversion.

Viral Regulatory Structures and Their Degeneracies, edited by Gerald Myers.
SFI Studies in the Sciences of Complexity, Proc. Vol. XXVIII, Addison-Wesley, 1998   **53**

The characteristic cell specificity of HPV infection has made *in vitro* studies difficult. An *in vitro* cell system has been available now for several years consisting of organotypic raft cultures of keratinocytes that allow complete differentiation of the epithelium *in vitro*.[1] However, despite the use of this system, many characteristics of the viral life cycle remain unknown. In particular, the close relationship between the viral vegetative cycle and the program of cell differentiation is not understood. Results from these systems have not given a clear picture of the events during the viral life cycle. We will discuss various points related to transcriptional regulation and expression of viral functions in these systems.

Cell lines established from cervical carcinomas have also been useful *in vitro* systems for the study of papillomavirus gene expression. Some of these cell lines are associated with HPV, while others are not. When HPV is associated with cervical carcinoma cell lines established from genital cancers, the viral DNA persists as an integrated copy(ies) in the cellular genome. Frequently, only part of the viral genome is found intact, containing the regulatory region and the E6 and E7 genes which are actively transcribed (McBride et al., this volume). Transcription of the E6 and E7 genes is under the control of a viral regulatory region in which lie the E6/E7 promoter and the keratinocyte-specific enhancer. Interestingly, the downstream open reading frames (ORFs) encoding the E1 and E2 regulatory proteins are usually disrupted.

We can speculate that part of the cell specificity of the papillomavirus infection is directly linked to transcriptional control. Transfection experiments in cell cultures have revealed several of the determinants that control this transcription and its cell specificity. The basal transcription from the E6/E7 promoter in the natural host, the keratinocyte, does not require viral proteins. In contrast, the viral E2 protein represses E6/E7 transcription. These two aspects of the E6 and E7 transcription relating to cell specificity and regulation by the viral transcription factor E2, is discussed in this chapter and in the preceding chapter by McBride and colleagues. We will discuss the relationship between this regulation and carcinogenic progression of HPV infection. In addition, since E2 is also involved in the viral DNA replication, we will discuss the interplay between transcription and replication controls in the HPV life cycle.

2. TISSUE SPECIFICITY

2.1 HPV TRANSCRIPTION IN ORGANOTYPIC CULTURES AND BENIGN LESIONS

Expression of the viral genes during the productive life cycle of human papillomavirus is tightly coupled to the differentiation program of epithelial cells. HPV can accomplish a complete cycle only in fully differentiated epithelia. In this regard, organotypic cultures, where keratinocytes can undergo complete differentiation, present unique systems in which to study the HPV life cycle. Organotypic raft cultures recreate important features of epithelial differentiation *in vitro* by raising the cells to an air/liquid interface. This has been done by growing epithelial cells on a collagen matrix for support.[1] There are different types of such cultures prepared from human foreskin inoculated with HPV (type 11), a type of genital papillomavirus not associated with carcinogenic progression, and from cells of low-grade lesions containing episomal HPV (type 31b).

In these cultures it has been shown that HPV DNA replication takes place in the upper layers of the epithelium, where the viral DNA is maintained in an episomal state.[2] However, production of virions could only be achieved by addition of 12-O-tetradecanoyl phorbol 13-acetate (TPA) to the medium. This treatment has been shown to induce better terminal differentiation of the epithelium *in vitro* and concomitantly allow production of HPV viral particles in the upper layers.[26] Two other systems have been reported that allow productive infectious cycle with HPV-11, human foreskin xenografts in nude mice[36] and a modified raft culture *in vitro*.[10]

Since the complete vegetative viral cycle was observed in these systems, one would expect to be able to study its various stages in the best "physiological" conditions *in vitro*. The viral oncoproteins E6 and E7 modulate cellular proliferation. Therefore, their expression would be expected in the basal layer containing the proliferating cells of the epithelium. The regulating viral proteins E1 and E2 should be expressed at later stages of the infection in more differentiated suprabasal layers preceding viral DNA amplification. The capsid proteins should only appear in the uppermost differentiated cells where virions are encapsidated and released into the environment. However, so far, this simple pattern has not been confirmed by studies of the viral transcription in infected raft cultures.[10] In the HPV-11-associated raft cultures, only low levels of messenger RNAs, encoding for E6 and E7, persisted in the terminally differentiated cell layers.[18] These data should be compared with studies performed by *in situ* hybridization on low-grade lesions, where E6 encoding mRNAs could not be detected while E7 encoding mRNAs were present at low abundance in basal and suprabasal layers.[20]

The viral E1 and E2 proteins are necessary for episomal replication of HPV genomes.[42] Yet, low or no mRNAs encoding these two proteins have been detected in raft cultures despite active viral DNA replication in the upper layers of the epithelium (Figure 1(b)). Therefore, these systems have not given a coherent view of the transcription pattern of HPV genomes during the vegetative cycle. Messenger RNAs for the E6 and E7 transforming proteins are not present in proliferative basal cells and those for E1 and E2 are hardly detectable in suprabasal layers, where viral genomes are actively replicated. In contrast, extremely abundant transcripts, which encode a protein resulting from the fusion of the five N-terminal amino acids of E1 coding region and the entire E4 ORF, are detected. These are initiated at a promoter within the E7 coding region that appears to be activated during cell differentiation.[10,18] The specific role of E4 protein has not been well defined in HPV infection. It seems to behave as a protein which can interact with cytokeratins and may help disrupt the cytoskeleton at the later stage of infection.[12,28] As expected however, abundant mRNAs encoding the capsid proteins L1 and L2 have been detected in the uppermost differentiated layers.

2.2 TRANSCRIPTION IN CERVICAL CARCINOMA CELL LINES

Studies with HPV-16- or HPV-18-associated cell lines have shown that viral DNA integration follows a common pattern in the majority of cases. Only part of the viral genome is integrated in the cellular genome with the concomitant disruption of the

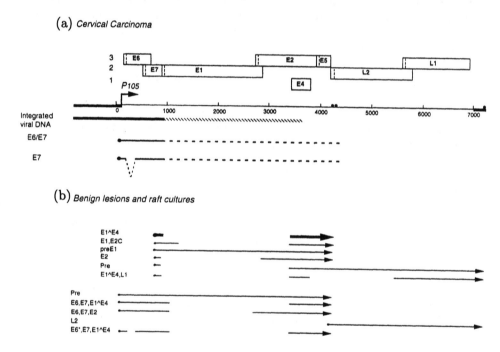

FIGURE 1   HPV transcription patterns. (a) HPV transcription in cervical carcinoma cell lines. The various open reading frames are presented as boxes in the three translation frames of the HPV-18 sequence. The complete HPV-18 genome of 7847 bp, is schematically represented as a continuous line with the regulatory region as a bold line. The portion of viral DNA integrated in cervical carcinoma cell lines, containing the regulatory region and the E6 and E7 genes, is schematized underneath. Black dots are the early (at around 4200) and late (at around 7000) polyadenylation signals. The two main messenger RNAs expressed in HPV-18-associated carcinoma cell lines are also shown.[33] (b) HPV transcription in benign lesions and in raft cultures *in vitro*: The structures of the various RNAs described from systems where productive viral cycle occurs *in vivo* as well as *in vitro* are shown and their coding potential indicated. The most abundant RNA, encoding the E1∧E4 gene product, is shown in bold line.[18,20,27,36]

E1 or E2 ORFs downstream of the E6 and E7 region. The integrated viral genome is actively transcribed as two major RNAs that encode for the viral proteins E6 and E7[33] (Figure 1(a)). These mRNAs have a chimeric structure containing cellular sequences at their 3′ end. Recently, such sequences have been shown to stabilize these mRNAs.[21] The E6 and E7 gene products are continuously expressed in genital tumors or cell lines established from them and their expression has been shown to be necessary to maintain the transformed phenotype.

Transcription of the viral mRNAs encoding for the transforming proteins initiates at single promoters contained in the integrated HPV-16 or HPV-18 regulatory regions (LCR) (Figure 1). They are named P_{97} and P_{105}, respectively, after the nucleotide at which the RNA is initiated. The structure and activity of these promoters have been extensively studied, and we will describe some aspects of their regulation, focusing on the regulation of the HPV-18 P_{105} promoter. Studies from our group and others have shown that HPV transcriptional regulation is strictly cell specific.

2.3 STRUCTURE OF THE REGULATORY REGION

The regulatory region in human papillomaviruses extends from the end of the late region to the beginning of the early region, over about 800 bp (Figure 2). These sequences are much less conserved among the different genital viruses than the protein coding regions. They contain signals for the control of transcription as well as viral DNA replication. As described earlier, this region is intact within the integrated part of HPV genomes in cervical carcinomas. It has been shown to contain a cell-specific enhancer of transcription, which is primarily active in epithelial cells of human origin.[15,16] Its activity depends, to a large extent, on cellular transcription factors. It is situated about 200 bp upstream of the P_{105} or P_{97} promoters in HPV type 18 or 16, respectively. The promoter proximal regions themselves (within the proximal 200 bp) do not exhibit any intrinsic transcriptional activity, despite the presence of binding sequences for cellular transcription factors.

Numerous cellular factors interact with the regulatory regions of HPV-18 or HPV-16, as shown by DNAseI footprinting experiments, as well as by gel shift assays (Figure 2) and Garcia-Carranca et al.[15] and Gloss et al.[16] Surprisingly, mostly ubiquitous cellular transcription factors have been shown to play a functional role in the transcriptional regulation of the HPVs: notably, NF1, Sp1, and AP1. The glucocorticoid receptor, Oct1, or YY1, has also been shown to interfere negatively with transcription of the early messengers.

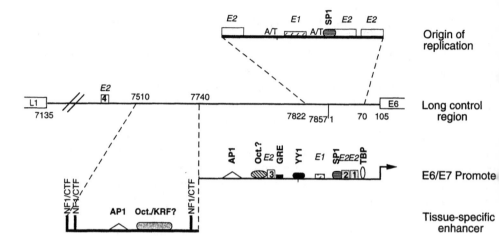

FIGURE 2   The HPV-18 regulatory region. Schematic representation of the 800 bp of the HPV-18 regulatory region. The tissue-specific enhancer and the origin of replication are shown. Binding sites for cellular and viral transcription factors are schematically represented along this sequence.

2.4 ROLE OF AP1 AND CELLULAR TRANSCRIPTION FACTORS IN TISSUE SPECIFICITY

HPV-18 transcription *in vitro* is strictly restricted to human epithelial cells of various origins, although the best recipients are cells established from cervical carcinomas and human keratinocytes. When the cell specificity of HPV transcription was first established *in vitro*, it seemed that its basis would be quickly resolved. It appears now that the mechanism(s) has been elusive and may imply a complex array of factors, none of which is in itself cell specific.

The cellular transcription factor AP1 is a heterodimer composed of members of the Jun and Fos multigenic families. We found that HPV-18 transcription is activated by AP1 through interaction of this factor with two binding sites in the viral regulatory region. Moreover, human keratinocytes contain high levels of AP1, with a marked preference for JunB compared to the other members of the Jun family, cJun and JunD.[40] By site-directed mutagenesis, we found that binding of AP1 to the HPV-18 regulatory region is necessary for the basal activity of the P_{105} promoter in keratinocytes, but is not sufficient. We, therefore, think that AP1 may contribute to the cell specificity of viral transcription, but only when interacting with other factor(s).

A way to search for keratinocyte-specific factors would be to reexamine the specificity of some of the so-called ubiquitous transcription factors that interact with HPV-18 such as Sp1, NF1/CTF, or Oct. These factors belong to multigenic families, members of which could be expressed differentially in different cell types or during keratinocyte differentiation. Another approach that could help elucidate HPV transcriptional regulation would be to analyze by immunofluorescence the presence of various cellular transcription factors in thin sections of lesions or normal skin and study their modification during keratinocyte differentiation and viral infection.

3. ROLE OF THE VIRAL E2 TRANSCRIPTIONAL REGULATOR

Since the initial observation in 1985 that the bovine papillomavirus type 1 (BPV1) E2 ORF encodes a transcriptional *trans*-activator,[34] HPVs E2 proteins have been shown to share this property. The full-length product of the E2 ORF is a protein of about 50 kDa which is composed of three distinct domains as predicted by structural and biochemical studies. This structure is well conserved among papillomaviruses despite their divergent primary sequences. (For example, the amino acid identity between the conserved *N*-terminal and the *C*-terminal domains of the BPV1 and HPV-18 E2 proteins is only about 30%, while between HPV type 18 and 16 it is about 50%.) The *N*-terminal part of E2 contains the transcriptional activation domain, while the *C*-terminal part contains the dimerization and DNA binding domain, (see Ham et al.[17] for review).

3.1 E2-MEDIATED TRANSACTIVATION

Transactivation by the E2 protein depends upon its binding as a dimer to a palindromic sequence ACCG(N4)CGGT, which is present in multiple copies in PV genomes. For example, there are 17 such sites in the BPV1 genome but only 4 in genital HPVs[25] (Figure 2). E2 can also activate transcription from heterologous promoters, such as those of the HSV (herpes simplex virus) thymidine kinase (tk) gene or the early region of SV40, in the presence of binding sites either upstream or downstream of the promoter sequences. While a single E2 palindrome activates transcription only weakly (maximum six fold), two E2 sites can function cooperatively as an E2 dependent enhancer (fifty to hundred fold activation). Most functional properties of E2 are shared by all E2 proteins studied so far. In particular, the E2 proteins of HPV types 11, 16, or 18 have been shown to activate transcription synergistically from heterologous promoters containing multiple E2 binding sites.

3.2 E2-MEDIATED REPRESSION

All papillomaviruses contain a transcriptional transactivator encoded by the E2 open reading frame and cognate DNA-binding sites in their regulatory regions. However, when studied in the context of transcription of genital human papillomaviruses, E2 represses rather than activates transcription from the E6/E7 genes.[29,41] This transcriptional repression involves binding of the full-length E2 protein to E2 binding sites close to the TATA box of the E6/E7 promoters (Figure 3). Thus, the mechanism by which the transcription is repressed appears to involve steric hindrance at the site of formation of the transcriptional initiation complex.[13]

The four E2 binding sites present in the regulatory regions of genital HPVs are conserved in their sequence, as well as their position relative to the TATA box in the E6/E7 promoter (Figure 2). Repression by E2 is relieved when mutations, that prevent binding of E2, are introduced in the two binding sites most proximal to the promoter TATA box (Figure 3).[39] The mechanism of this transcriptional repression, which occurs even with the full-length E2 protein, is distinct from that of the repression of the BPV1 P_{89} promoter whose E2-mediated activation is repressed by competitive formation of E2 heterodimers between a full-length protein and a truncated form containing only the C-terminal, DNA-binding, and dimerization domain. However, it may resemble the mechanism of repression of the P_1 promoter, in BPV1, which occurs by competitive binding of E2 with a cellular factor, reviewed in Ham et al.[17] In the case of HPVs, two putative factors may be involved in this competitive binding by E2. First, the TATA box binding protein (TBP), since the most proximal E2 binding site in HPVs is situated only 3 or 4 nt upstream of

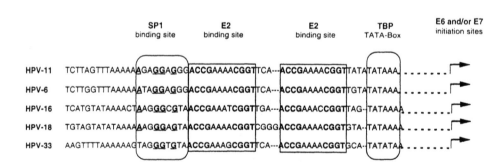

FIGURE 3 Sequence comparison between the E6/E7 promoter proximal regions of HPVs. Sequence comparison between the proximal sequences of the E6/E7 promoters of five different HPVs. The conserved TATA boxes, Sp1 binding sites and the E2 binding sites are shaded. The nucleotides conserved among Sp1 binding sites in HPVs are shown in bold and underlined. Spacing between the different boxes is rather well conserved among genital HPVs as indicated.

the TATA box of the E6/E7 promoter (Figure 3). *In vitro* binding experiments have indicated that binding of E2 may negatively interfere with the binding of TBP on the proximal TATA box in the HPV-18 P_{105}.[13] Second, SP1, since binding of E2 to the second site, which is further upstream of the TATA box sequence, interferes with the binding of this factor, essential for the activity of the early promoters (Figure 3). This exclusion of Sp1 by E2 appears to be either involved in the E2-mediated repression of the HPV early promoters in C33 cells[11,38] or not.[8]

At this point, it should be noted that the exact mechanism by which E2 represses the P_{105} and P_{97} promoters is not yet fully understood. The three most proximal E2 binding sites of the HPV-18 P_{105} promoter appear to be involved in the E2-mediated transcriptional repression depending on the cellular environment. When the three proximal E2 binding sites are mutated, E2 becomes a weak transcriptional *trans*-activator of P_{105}, presumably by interacting with the fourth most upstream E2 binding site. In addition, different forms of the E2 proteins can repress transcription differently, although they bind DNA similarly.[6] We should recall at this point that two reports claim that the E2 protein would activate transcription from the HPV-16 and HPV-18 early promoters, when present at low doses, while it represses transcription at high doses, the underlying putative mechanism being that E2 binds the various sites in the LCR with different affinities.[3,35] Since molecular mechanisms were only dissected for E2-mediated repression, it is impossible at the moment to draw a clear picture of the time course of E2-mediated regulation in HPV-infected lesions. All data mentioned above have been obtained by transient cotransfection experiments *in vitro*. We feel that different systems, such as inducible systems which allow more subtle regulation of E2 levels, will now have to be used to clarify this issue at the molecular level.

However, these data indicate that the crucial transcriptional regulatory function of the HPV E2 proteins, when bound to the three proximal binding sites of the P_{105} promoter, is to repress transcription of the E6 and E7 transforming genes of HPVs infecting the genital tract. This function requires only the dimerization and DNA binding domains of the E2 protein. In this regard, the role of the conserved transactivation domain in the transcription of genital papillomaviruses genes remains unclear. One could speculate that promoters, that are not active in transfected cells, are activated during the viral cycle. In BPV1 for instance, there are seven promoters for early or late gene expression. They are differentially regulated by E2, some being activated and others repressed. The epidermodysplasia verruciformis (EV)-associated cutaneous papillomavirus type 8, contains a promoter for late transcription which is activated by E2 while the early promoter is repressed.[37] Therefore, one simple explanation could be that the genital HPV E2 proteins activate transcription of late promoter(s) (yet unknown) and that this regulation could not be studied in the assays currently used which do not allow correct terminal differentiation of cells *in vitro* (late HPV transcription is strictly dependent upon keratinocyte terminal differentiation). In addition, the full-length E2 protein with both intact transactivation and DNA binding domains is required for the control of viral DNA replication as discussed below.

3.3 E2-MEDIATED GROWTH ARREST AND APOPTOSIS OF HeLa CELLS

One of the "physiological" consequences of the E2-mediated transcriptional repression of the E6 and E7 oncogenes emerged from studies of the HeLa cell line derived from HPV-18-associated cervical carcinoma, transfected with E2. These cells originally do not express E2 due to disruption of the E1/E2 region of the viral genome upon integration into the cellular genome, but efficiently express E6 and E7 as discussed earlier (Figure 1). We and others have found that reintroduction of E2 in these cells induced a growth arrest in G1.[9,14,19] This growth arrest appeared linked to the repression of the endogenous E6/E7 transcription by E2 and subsequent induction of p53. Therefore, the E2 protein can counteract the proliferative functions of both E6 and E7 in cervical carcinoma cell lines associated with HPV. In addition, we found that transfection of E2 into HeLa cells induced cell death by apoptosis. Both phenomena seemed linked to p53 induction, although p53 transcriptional activity appears not to be required for apoptosis (see Figure 4). We found that the induction of p53 by E2 in HeLa cells was mediated by two different mechanisms. One involves binding of E2 to cognate sequences in the E6/E7 promoter, leading to repression of endogenous E6 transcription. Repression of E6 transcription would allow an accumulation of the p53 protein by reducing its degradation by the E6-mediated ubiquitination pathway (30). However, a second pathway was unravelled by the ability of a mutated E2 protein, to increase the level of p53 activity, although this mutated E2 could not bind DNA and therefore could not repress E6 transcription. We conclude from our experiments that E2 carries two independent functions that cooperate in activating p53 in HeLa cells. One is binding to cognate sites in the E6/E7 promoter, leading to repression of the endogenous E6 transcription, while the other is independent of DNA binding and requires an intact N-terminal domain (Figure 4).

In conclusion, these experiments unravelled a new role for E2 in controlling both growth and apoptosis of HPV-transformed cells. This negative control of cell proliferation by E2 provided a satisfactory explanation for the fact that the E2 ORF is preferentially disrupted, or its expression blocked, in cervical cancers associated with HPV-18. However, we do not know whether this putative function of E2 plays any role in the viral reproductive cycle. One could speculate that E2 would counteract the proliferative effect of E6 and E7 providing a tight and sensitive control of the infected cells in the basal layer of the epithelium. On the other hand, in the upper layers, where virions are assembled, expression of E2 could participate in the induction of apoptosis therefore allowing the release of the mature particles into the environment.

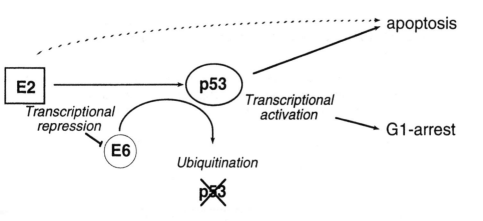

FIGURE 4   Model of E2-mediated G1 growth arrest and apoptosis in HeLa cells. E2 activates p53 by two different pathways, one requires repression of the E6 transcription and the other does not require this repression but is mediated through the N-terminal transactivation domain. Accumulation of p53 in HeLa cells induces G1-growth arrest and apoptosis, although apoptosis does not require transcriptionally active p53.[9]

4. TRANSCRIPTIONAL REGULATION AND VIRAL DNA REPLICATION

4.1 FUNCTIONAL ROLES OF THE E1 AND E2 PROTEINS IN REPLICATION

Normally, in infected cells, papillomaviral genomes exist as minichromosomes with a chromatin structure identical to that of the cellular DNA. Viral DNA replication takes place in the nucleus of infected cells and is dependent on the host cell's nuclear machinery. The main viral replication protein is E1 which is the longest well-conserved, early viral gene product among papillomaviruses. Two regions could be distinguished within its primary sequence, a highly conserved C-terminal domain containing various sequences with homology to SV40 T Ag such as ATP binding, ATPase and helicase domains, and a divergent N-terminal domain.[4] However, papillomavirus DNA replication optimally requires the full-length viral transcriptional regulator E2. The minimal origin of replication contains binding sites for E2 and E1, and E1 alone cannot bind DNA efficiently (Figure 2). Therefore, it appears that E1 and E2 form a complex that binds to the DNA, the main role of E2 being to target E1 to this origin.

5. VIRAL DNA REPLICATION AND TRANSCRIPTIONAL REGULATION

Regulatory regions of HPVs contain elements controlling viral DNA replication that overlap with the proximal sequences of the P_{105} promoter as shown in Figure 2. The origin of replication is contained within a 100 bp fragment including three E2 binding sites surrounding an A/T rich region in which lies the E1 binding site.[5,7] Interestingly, this region also contains binding sites for the cellular transcription factors Sp1 and YY1 and is closely bordered by the TATA sequence, which binds TBP (Figure 2).

5.1 ROLE OF CELLULAR TRANSCRIPTION FACTORS IN VIRAL DNA REPLICATION

Transcription factors have been shown to play auxiliary roles in activating the replication of SV40 and polyomavirus DNA. The situation encountered with papillomaviruses is somewhat different in that viral DNA replication requires, in addition to E1, the E2 transcription factor. The function of transcription factors in the activation of viral DNA replication may, therefore, be sustained by E2 with papillomavirus DNA. Some of these functions have been suggested for the BPV1 E2 such as binding of RPA (replication protein A) or modification of the chromatin structure around the origin of replication.[23,24] Cellular transcription factors, whose binding sites are present either in the origin of replication or close to it, also play an auxiliary activating function in HPV DNA replication, although their role remains minor when compared to E2.[5,30]

5.2 THE REPLICATION INITIATION COMPLEX CONTAINING E2 AND E1 ACTIVATES TRANSCRIPTION

Direct physical interaction between the viral E1 and E2 proteins have been demonstrated. In addition to playing a functional role in the control of replication, this interaction could also interfere with the E2-mediated activation of transcription. In the BPV1 system, we have shown that E2 and E1 cooperatively activate transcription of a minimal promoter (HSV1 tk), harboring upstream the BPV1 origin of replication. This synergistic activation depends upon cooperative binding of the two proteins on two adjacent binding sites in the origin of replication.[22]

These results indicate that E2 remains active when engaged in the E1/E2 initiation complex on the viral origin of replication. Whether such transcriptional activation plays any role in the viral DNA replication or during the viral cycle is unknown. On the other hand, excess E1 has been shown to repress E2-mediated transcriptional activation.[22,31] During infection, accumulation of E2 and E1 will result in the repression of E6/E7 transcription by the binding of E2 to sites 1, 2, and

3 and the concomitant activation of replication by assembly of the replication initiation complex containing E1 and E2 at the origin. In this respect, the interplay between transcription and replication resembles that operating with polyomaviruses. Binding of TAg near the origin represses early transcription and activates DNA replication. However, in this case, a single viral protein participates in this switch while, with papillomaviruses, two viral proteins are essential for replication and only E2 is sufficient for repression.

6. CONCLUSION

Studies of the regulation of HPV transcription have so far only concerned the regulation of a single promoter. This promoter has proven to be of extreme importance, since it regulates the initiation of the two viral RNAs encoding the oncoproteins E6 and E7. In addition, it is the only viral promoter active in cell lines established from cervical carcinomas associated with HPV types 16 or 18. Its expression is cell-type specific and occurs only in epithelial cells of human origin. In these cells, the HPV-16 P_{97} or HPV-18 P_{105} promoters have been shown to be activated by a complex array of cellular factors, although some negative regulation has also been reported.

In this context, the main viral transcriptional regulator, the product of the E2 ORF, represses transcription. This repression occurs by binding of the full-length protein to one or two of the E2-binding sites immediately adjacent to the TATA box of the promoter. Disruption of the E1 or E2 open reading frame by integration of the viral DNA into the cellular genome, which occurs frequently in cervical cancers associated with HPV type 18, would abolish E2 expression. It would therefore lead to derepression of transcription at the P_{105} promoter which, in turn, would allow enhanced expression of the E6 and E7 oncoproteins in the cells. Further support for this hypothesis comes from the demonstration that the introduction of E2 into a cervical carcinoma cell line associated with HPV-18 can counteract the proliferative functions of the viral oncogenes, inducing growth arrest and apoptosis, at least partly by repressing their endogenous transcription. It should be noted that the disruption of the E1 and E2 open reading frames also prevents viral DNA replication.

One of the most intriguing aspects of the HPV vegetative cycle is the relationship existing between the control of transcription and viral DNA replication, as well as the respective roles of the two viral regulators E2 and E1. Another open question is the role of the E2 *trans*-activator in the "normal" viral cycle. A crucial step in the study of the molecular mechanisms regulating HPV transcription and replication would be the setting up of an *in vitro* cell system that would allow both keratinocyte differentiation and analysis of viral gene expression to take place during the vegetative reproductive cycle.

66F. Thierry and M. Yaniv

ACKNOWLEDGMENTS

We wish to thank Caroline Demeret and Christian Desaintes for helpful discussions and suggestions and Caroline Demeret and Jonathan Weitzman for critical reading of the manuscript.

REFERENCES

1. Asselineau, D., B. Bernard, C. Bailly, and M. Darmon. "Epidermal Morphogenesis and Induction of the 67 kD Keratin Polypeptide by Culture of Human Keratinocytes at the Liquid-Air Interface." *Exp. Cell Res.* **159** (1985): 536–539.

2. Bedell, M. A., J. B. Hudson, T. R. Golub, M. E. Turyk, M. Hosken, G. D. Wilbanks, and L. A. Laimins. "Amplification of Human Papillomavirus Genomes *In Vitro* is Dependent on Epithelial Differentiation." *J. Virol.* **65** (1991): 2254–2260.

3. Bouvard, V., A. Storey, D. Pim, and L. Banks. "Characterization of the Human Papillomavirus E2 Protein: Evidence of Trans-Activation and Trans-Repression in Cervical Keratinocytes." *EMBO J.* **13** (1994): 5451–5459.

4. Clertant, P., and I. Seif. "A Common Function for Polyoma Virus Large-T and Papillomavirus E1 Proteins?" *Nature* **311** (1984): 276–279.

5. Demeret, C., M. L. Moal, M. Yaniv, and F. Thierry. "Control of HPV-18 DNA Replication by Cellular and Viral Transcription Factors" *Nucl. Acids Res.* **23** (1995): 4777–4784.

6. Demeret, C., C. Desaintes, M. Yaniv, and F. Thierry. "Different Mechanisms Contribute to the E2-Mediated Transcriptional Repression of HPV-18 Viral Oncogenes." *J. Virol.* (1997): in press.

7. Demeret, C. Personal communication. Ph.D. Thesis, Paris 6, Nov. 1996.

8. Demeret, C., M. Yaniv, and F. Thierry. "The E2 Transcriptional Repressor Can Compensate for Sp1 Activation of the Human Papillomavirus Type 18 Early Promoter." *J. Virol.* **68** (1994): 7075–7082.

9. Desaintes, C., C. Demeret, S. Goyat, M. Yaniv, and F. Thierry. "Expression of the Papillomavirus E2 Protein in HeLa Cells Leads to Apoptosis." *EMBO J.* (1997): in press.

10. Broker, T. R., and L. T. Chow. "Production of Human Papillomavirus and Modulation of the Infectious Program in Epithelial Raft Cultures." *Genes & Dev.* **6** (1992): 1131–1142.

11. Dong, G., T. R. Broker, and L. T. Chow. "Human Papillomavirus Type 11 E2 Proteins Repress the Homologous E6 Promoter by Interfering with the Binding of Host Transcription Factors to Adjacent Elements." *J. Virol.* **68** (1994): 1115–1127.

12. Doorbar, J., S. Ely, J. Sterling, C. McLean, and L. Crawford. "Specific Interaction Between HPV-16 E1-E4 and Cytokeratins Results in Collapse of the Epithelial Cell Intermediate Filament Network." *Nature* **352** (1991): 824–827.

13. Dostatni, N., P. F. Lambert, R. Sousa, J. Ham, P. M. Howley, and M. Yaniv. "The Functional Bpv1 E2 Trans-Activating Protein Can Act as a Repressor by Preventing Formation of The Initiation Complex." *Genes & Dev.* **5** (1991): 1657–1671.

14. Dowhanick, J. J., A. A. McBride, and P. M. Howley. "Suppression of Cellular Proliferation by the Papillomavirus E2 Protein." *J. Virol.* **69** (1995): 7791–7799.
15. Garcia-Carranca, A., F. Thierry, and M. Yaniv. "Interplay of Viral and Cellular Proteins Along the Long Control Region of Human Papillomavirus Type 18." *J. Virol.* **62** (1988): 4321–4330.
16. Gloss, B., H. U. Bernard, K. Seedorf, and G. Klock. "The Upstream Regulatory Region of the Human Papillomavirus −16 Contains an E2 Protein Independent Enhancer which Is Specific for Cervical Carcinoma Cells and Regulated by Glucocorticoid Hormones." *EMBO J.* **6** (1987): 3735–3743.
17. Ham, J., N. Dostatni, J. M. Gauthier, and M. Yaniv. "The Papillomavirus E2 Protein: A Factor with Many Talents." *Trends Biochem. Sci.* **16** (1991): 440–444.
18. Hummel, M., J. B. Hudson, and L. A. Laimins. "Differentiation-Induced and Constitutive Transcription of Human Papillomavirus Type 31b in Cell Lines Containing Viral Episomes." *J. Virol.* **66** (1992): 6070–6080.
19. Hwang, E. S., D. J. Riese, J. Settleman, L. A. Nilson, J. Honig, S. Flynn, and D. D. Maio. "Inhibition of Cervical Carcinoma Cell Line Proliferation by the Introduction of a Bovine Regulatory Gene." *J. Virol.* **67** (1993): 3720–3729.
20. Ifner, T., M. Oft, S. Bohm, S. P. Wilczynski, and H. Pfister. "Transcription of the E6 and E7 Genes of Human Papillomavirus Type 6 Anogenital Condyloma Is Restricted to Undifferentiated Cell Layers of the Epithelium." *J. Virol.* **66** (1992): 4639–4646.
21. Jeon, S., and P. F. Lambert. "Integration of Human Papillomavirus Type 16 DNA into the Human Genome Leads to Increased Stability of E6 and E7 mRNAs: Implications for Cervical Carcinogenesis." *Proc. Natl. Acad. Sci. USA* **92** (1995): 1654–1658
22. Le Moal, M. A., M. Yaniv, and F. Thierry. "The Bovine Papillomavirus Type 1 (BPV1) Replication Protein E1 Modulates Transcriptional Activation by Interacting with BPV1 E2." *J. Virol.* **68** (1994): 1085–1093.
23. Levebvre, O., G. Steger, and M. Yaniv. "Synergistic Transcriptional-Activation by the Papillomavirus E2 Protein Occurs after DNA Binding and Correlates with a Change in Chromatin Structure." *J. Mol. Biol.* **267** (1997): in press.
24. Li, R., and M. R. Botchan. "Acidic Transcription Factors Alleviate Nucleosome-Mediated Repression of DNA Replication of Bovine Papillomavirus Type 1." *Proc. Natl. Acad. Sci. USA* **91** (1994): 7051–7055.
25. Li, R., J. Knight, G. Bream, A. Stenlund, and M. Botchan. "Specific Recognition Nucleotides and Their DNA Context Determine the Affinity of E2 Protein for 17 Binding Sites in the BPV-1 Genome." *Genes Dev* **3** (1989): 510–526.
26. Meyers, C., M. G. Frattini, J. B. Hudson, and L. A. Laimins. "Biosynthesis of Human Papillomavirus From a Continuous Cell Line Upon Epithelial Differentiation." *Science* **257** (1992): 971–973.

27. Pray, T. R., and L. A. Laimins. "Differentiation-Dependent Expression of E1–E4 Proteins in Cell Lines Maintaining Episomes of Human Papillomavirus Type 31b." *Virology* **206** (1995): 679–685.
28. Rogel-Gaillard, C., F. Breidburd, and G. Orth. "Human Papillomavirus Type 1 E4 Proteins Differing by Their *N*-Terminal Ends Have Distinct Cellular Localisations when Transiently Expressed *In Vitro*." *J. Virol.* **66** (1992): 816–823.
29. Romanczuk, H., F. Thierry, and P. M. Howley. "Mutational Analysis of *Cis* Elements Involved in E2 Modulation of Human Papillomavirus Type 16 P97 and Type 18 P_{105} Promoters." *J. Virol.* **64** (1990): 2849–2859.
30. Russell, J., and M. Botchan. "*Cis*-Acting Components of Human Papillomavirus (HPV) DNA Replication: Linker Substitution Analysis of the HPV Type 11 Origin." *J. Virol.* **69** (1995): 651–660.
31. Sandler, A. B., S. B. Vande Pol, and B. A. Spalholz. "Repression of Bovine Papillomavirus Type 1 Transcription by the E1 Replication Protein." *J. Virol.* **67** (1993): 5079–5087.
32. Scheffner, M., J. M. Huibregtse, R. D. Vierstra, and P. M. Howley. "The HPV-16 E6 and E6-AP Complex Functions as Ubiquitin-Protein Ligase in the Ubiquitination of p53." *Cell* **75** (1993): 495–505.
33. Schwarz, E., U. K. Freese, L. Gissmann, W. Mayer, B. Roggenbuck, A. Stremlau, and H. zur Hausen. "Structure and Transcription of Human Papillomavirus Sequences in Cervical Carcinoma Cells." *Nature* **314** (1985): 111–114.
34. Spalholz, B. A., Y. C. Yang, and P. M. Howley. "Transactivation of a Bovine Papillomavirus Transcriptional Regulatory Element by the E2 Gene Product." *Cell* **42** (1985): 183–191.
35. Steger, G., and S. Corbach. "Dose-Dependent Regulation of the Early Promoter of Human Papillomavirus Type 18 by the Viral E2 Protein." *J. Virol.* **71** (1997): 50–58.
36. Stoler, M. H., A. Whitbeck, S. M. Wolinsky, T. R. Broker, L. T. Chow, M. K. Howett, and J. W. Kreider. "Infectious Cycle of Human Papillomavirus Type 11 in Human Foreskin Xenografts in Nude Mice." *J. Virol.* **64** (1990): 3310–3318.
37. Stubenrauch, F., J. Malejczyk, P. G. Fuchs, and H. Pfister. "Late Promoter of Human Papillomavirus Type 8 and Its Regulation." *J. Virol.* **66** (1992): 3485–3493.
38. Tan, S. H., L. E. C. Leong, P. A. Walker, and H. U. Bernard. "The Human Papillomavirus Type 16 E2 Transcription Factor Binds with Low Cooperativity to Two Flanking Sites and Represses the E6 Promoter Through Displacement of Sp1 and TFIID." *J. Virol.* **68** (1994): 6411–6420.
39. Thierry, F., and P. M. Howley. "Functional Analysis of E2-Mediated Repression of the HPV-18 P_{105} Promoter. *New Biol.* **3** (1991): 90–100.

40. Thierry, F., G. Spyrou, M. Yaniv, and P. M. Howley. "Two AP1 Sites Binding JunB Are Essential for HPV-18 Transcription in Keratinocytes." *J. Virol* **66** (1992): 3740–3748.

41. Thierry, F., and M. Yaniv. "The BPV1-E2 Trans-Acting Protein Can Be Either an Activator or a Repressor of the HPV-18 Regulatory Region." *EMBO J.* **6** (1987): 3391–3397.

42. Ustav, M., and A. Stenlund. "Transient Replication of BPV-1 Requires Two Viral Polypeptides Encoded by the E1 and E2 Open Reading Frames." *EMBO J.* **10** (1991): 449–457.

Yesu Chao, Thomas P. Cujec, Jon K. Meyer, Hiroshi Okamoto, and B. Matija Peterlin*

Howard Hughes Medical Institute, Departments of Medicine, Microbiology, and Immunology, University of California–San Francisco, San Francisco, CA 94143–0724;
*E-mail: Matija@ITSA.UCSF.EDU

Transcriptional Regulation of HIV

All retroviruses have an obligatory intracellular phase where upon entry into cells they uncoat their virions, transcribe their RNA to DNA, transport reverse transcriptase complexes from the cytoplasm to the nucleus, integrate proviruses into the host genome, transcribe their genes, export different transcripts to the cytoplasm, translate viral proteins, assemble new virion particles, and infect other cells. As proviruses, they behave like any cellular gene, where the 5' LTR is the site of initiation of transcription and the 3' LTR contains polyadenylation and termination signals. This chapter focuses on cellular regulatory proteins that bind to the HIV-1 LTR to initiate transcription and on the viral *trans*-activator Tat that binds to the transactivation response region (TAR) to elongate viral transcription.

Viral Regulatory Structures and Their Degeneracies, edited by Gerald Myers.
SFI Studies in the Sciences of Complexity, Proc. Vol. XXVIII, Addison-Wesley, 1998

THE HIV-1 LONG TERMINAL REPEAT (LTR)

The 5′ LTR consists of U3, R, and U5 sequences. The site of initiation of transcription is at the U3/R junction (Figure 1). More than a dozen proteins bind to this region.[13] Of these, the TATA-binding protein (TBP) and the initiator are required for the positioning of the start site. Sp1 attracts more preinitiation complexes to the promoter. NFκB, LEF, Ets1, and NFAT bind to the 5′ enhancer sequences and increase not only rates of initiation, but also elongation, of viral transcription. These proteins transmit cellular signals to DNA: NFκB and NFAT respond to signals via the T cell antigen receptor, costimulatory signals, lymphokines, and cytokines. They are responsible for the synthesis of sufficient viral transcripts that code for Tat (i.e., for breaking molecular latency) and, together with Tat, for high levels of viral transcription and replication.[13] NFκB and NFAT are preassembled in the cytoplasm. Whereas NFκB must be released from IκB for its cytoplasmic to nuclear translocation, NFAT undergoes a conformational change to achieve the same goal. When bound to the LTR, they act synergistically to increase viral transcription.[13] Effects of these proteins are more pronounced with DNA templates that are covered with histones, i.e., in chromatin (Figure 1).[19,22]

Several regulatory proteins also bind to R and U5 sequences, such as LBP, CBP, AP1, NFAT, and others.[25] It is possible that a combination of upstream and downstream factors are required to keep the LTR in an open, i.e., accessible conformation.[24] Additionally, some of these proteins might induce the short transcripts (IST) that are transcribed from the LTR in the absence of Tat.[21] A signature of primate lentiviruses is that in their basal state, promoter proximal transcripts that contain TAR are synthesized.[13]

The relevance of some of these regulatory elements probably should be reevaluated in light of experiments with SIV, where proviruses containing no NFκB or Sp1 binding sites still replicated to wild-type levels and led to AIDS in rhesus macaques.[12] Whether other upstream or downstream elements from SIV or neighboring cellular genes substituted for these promoter proximal sequences is unknown.

Regulatory proteins encoded by several other DNA viruses can also activate transcription from the LTR. For example, early gene products from CMV, EBV, HBV, HHV6, HSV, HTLV, etc. can increase transcription from the LTR via different promoter elements (Figure 1).[13] However, the importance of these other regulatory proteins for HIV replication is unclear. Most of the time, HIV and these other viruses are found in different cells and these *trans*-activators are not able to cross cellular membranes to affect the transcription of genes in neighboring cells.

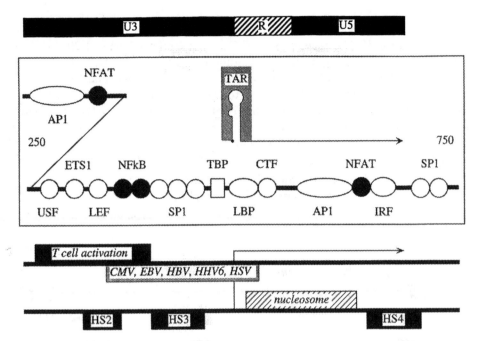

FIGURE 1   The HIV-1 LTR. Below the representation of the U3, R, and U5 sequences in the 5' LTR are transacting factors that interact with these sequences. Far upstream in the transcriptional enhancer are binding sites for AP1, NFAT, USF, ETS1, LEF, and NFκB. The core promoter binds Sp1, TBP, Initiator, LBP, and CTF. In the 3' untranslated region are binding sites for AP1, NFAT, IRF, and Sp1. Regulatory proteins that bind to the LTR only after cellular activation are depicted as solid (black) circles. The site of initiation of transcription is at the 5' border of sequences occupied by the Initiator and LBP. TAR is formed at the 5' end of all viral transcripts. Below the diagram of the LTR are depicted regions which respond to T cell activation and early *trans*-activators of a number of other human viruses. Below that diagram are presented three hypersensitive sites and the major nucleosome present in transcriptionally silent LTR. In transcriptionally active LTR, the entire region depicted in this figure is free of nucleosomes.

TRANS-ACTIVATOR TAT

The HIV *trans*-activator Tat is one of the earliest viral proteins made from the integrated provirus.[13] Four different multiply spliced transcripts encode Tat (Figure 2). Whereas three of them contain two exons of Tat, the fourth transcript synthesizes a molecule called Tev, which is short for Tat, Env, and Rev. This protein represents a chimera encoded by exon 1 of *tat*, the middle portion of *env* and the second exon of *rev*. Surprisingly, Tev maintains significant Tat and Rev activities.[4] Exon

1 of Tat encodes a protein of 72 amino acids, which are sufficient for optimal levels of transactivation.[13] The second exon of Tat encodes an additional 20–40 amino acids, which have been implicated in the differential transactivation of integrated versus transfected proviral DNA, for facilitated entry of extracellular Tat into cells via the integrin pathway, and for leading to the activation and/or apoptosis of infected or bystander T cells. Its N-terminal 72 amino acids can be further subdivided into N-terminal, cysteine-rich, core, basic, and C-terminal domains (Figure 3). Of these sequences, the cysteine-rich, core, and basic domains represent the essential module of Tat, of which the cysteine-rich and core domains are required for transcriptional activation and the basic domain for binding to RNA and targeting Tat to the nucleus.[13] The basic domain of 9 amino acids, which is rich in arginines and lysines, binds to the 5′ bulge in TAR (Figure 4).[7,15]

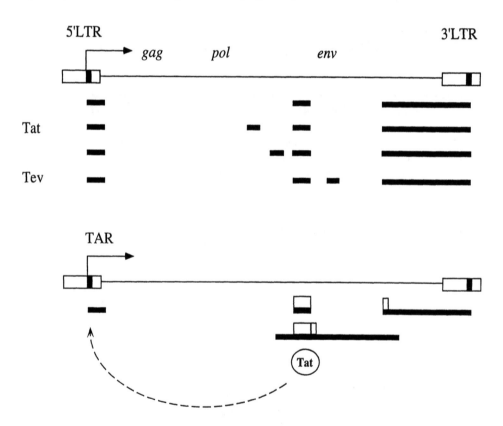

FIGURE 2   The biosynthesis of Tat. Tat is synthesized from four different multiply spliced HIV transcripts. Its two coding exons flank the *env* gene. The fourth transcript contains the first exon of *tat*, the middle of *env*, and the second exon of *rev*. The prototypic Tat is synthesized from 2 exons and then binds to TAR in the 5′ LTR to increase rates of elongation of viral transcription.

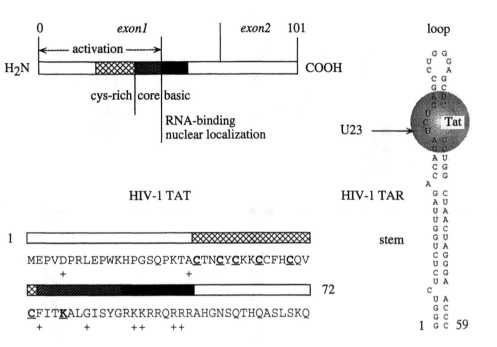

FIGURE 3 The structure of Tat. Tat contains between 82–101 amino acids, 72 of which are encoded in its first exon. Five functional domains have been defined. N- and C-terminal domains flank the cysteine-rich, core, and basic domains of Tat. The cysteine-rich and core sequences represent the activation domain of Tat and the basic domain binds to the major groove in the 5' bulge of TAR (see the right hand panel). Residues of the first exon of Tat are given, with those essential for Tat transactivation represented by bold, underlined letters. "+" signs represent amino acids whose mutations also have significant effects on the activity of Tat. Tat binds to TAR, which forms a 59-nucleotide RNA stem-loop. Tat binds to the 5' bulge in TAR, where the bulged U23 is most important. Other cellular proteins bind to the loop in TAR.

Functional domains of Tat have been investigated using heterologous RNA- and DNA-tethering proteins. For example, using the coat protein of bacteriophage MS2 and its operator RNA target, or Rev and its stem loop 2B (SLIIB) from the Rev response element (RRE) RNA target, the N-terminal 48 amino acids of Tat function as an independent activation domain.[13] Although less potent, these 48 amino acids can also activate transcription via DNA as Gal 4 or Lex A fusion proteins.[13] However, the mechanism of action of Tat via DNA and RNA is not equivalent.[20] The 9 amino acid basic domain is sufficient to bind to TAR in the nanomolar range, but flanking amino acids potentiate these protein-RNA interactions

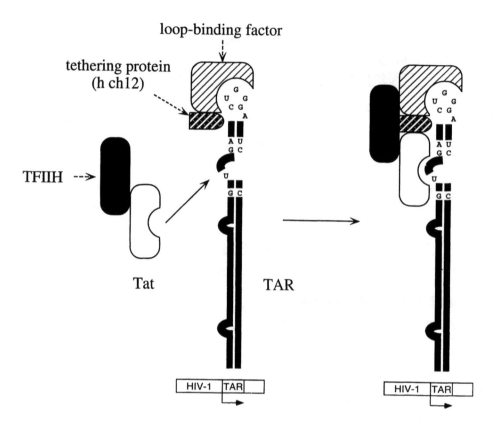

FIGURE 4   Productive interactions between Tat and TAR. Tat binds to the 5′ bulge in TAR with the help of cellular loop-binding proteins and a tethering factor encoded by human chromosome 12. The secondary structure of TAR is very important for these protein-RNA interactions and so are specific nucleotides represented by letters. Tat binds to and activates TFIIH via its activation domain.

and render them more specific.[7,15] Nevertheless, for optimal RNA-binding in cells additional proteins are required to presumably interact with the loop in TAR.[13]

The structure of Tat has been solved by NMR.[3] Tat is a largely unstructured protein that binds to the major groove of RNA.[8,9] Under physiological conditions, there are no α-helices in Tat.[3] Both N- and C-termini of Tat fold upon the core domain such that the molecule appears as a snake with its head and tail apposed to each other on its belly.[3] Since its basic domain has to fit into the major grove of the 5′ bulge in TAR, it is possible that in the presence of RNA, Tat undergoes a structural change. Additional structural constraints might be placed on Tat by different divalent cations that bind to its cysteine-rich domain and its cellular-interacting proteins. For the purpose of structural and functional analyses, a minimal lentiviral Tat was constructed from the activation domain of EIAV Tat (15 amino acids) and the basic domain from HIV Tat (10 amino acids).[5] The resulting 25-amino-acid Tat was able to transactivate the HIV-1 LTR via TAR to about 10% of wild-type levels. The creation of a minimal lentiviral Tat also reinforces the notion that Tat is a modular molecule which consists of discrete and well-defined functional domains.[17]

The availability of many different Tats from different lentiviruses has facilitated the study of proteins that interact with Tat. To date, almost 20 candidate proteins have been isolated, among them basal transcription factors such as TBP, TFIIB, TFIIF, TFIIH, and a CTD kinase, which might be distinct from TFIIH.[13,28] Additionally, transcription factors such as Sp1, p32, CTD phosphatase, a protein of 140 kDa called TatSF1, HIITA etc., have been proposed as specific co-activators of Tat.[13,30] However, for most of them, whether isolated by genetic or biochemical approaches, convincing functional correlates are lacking. Only TatSF1 has been demonstrated to be important for effects of Tat in an *in vitro* transcription system and to a lesser degree *in vivo*.[30] Moreover, the interaction between Tat and TFIIH is intriguing because it results in increased activity of the TFIIH kinase to phosphorylate the C-terminal domain (CTD) of RNA polymerase II.[18]

TRANSACTIVATION RESPONSE ELEMENT TAR

Tat is unique among eukaryotic transcriptional regulators in that it exerts its effects via an RNA target. TAR of HIV-1 contains a 59 nucleotide stem-loop with the predicted free energy of 37 kilocalories that has a prominent 5' bulge and a hexanucleotide loop which is rich in G residues. As mentioned previously, Tat interacts with the 5' bulge in TAR as a nascent RNA structure cotranscriptionally, and the formation of TAR is only required during the early stages of transcription at which point Tat acts upon the RNA polymerase II.[13] The notion that Tat interacts with TAR RNA and not TAR DNA was first established by extensive mutagenesis and compensatory mutations in TAR, by the replacement of TAR with the operator of bacteriophage MS2 or SLIIB from the RRE, and the use of the fusion proteins between Tat and the coat protein or Rev. In all these instances, Tat activated viral transcription via these heterologous RNA targets.[13] In the absence of Tat, all lentiviral promoters express TAR, which can be recovered as stable RNA in transfected and infected cells.[13] Overexpressed TAR also represents TAR decoys, which can be used to remove Tat from TAR and thus block Tat transactivation in cells.[23] As a form of gene therapy, TAR decoys could represent a component of antiretroviral therapy.

Tat binds to the major groove of the 5' bulge in TAR with a high affinity and low specificity.[6,15] The loop in TAR probably forms a more rigid structure.[1] There is an obligatory fixed distance between the 5' bulge and loop that cannot be varied, which implies that loop-binding proteins are essential for productive interactions between Tat and TAR.[6,15] This observation has been confirmed *in vitro* and *in vivo.* Moreover, several TAR loop-binding proteins have been isolated and cloned, chief among them the TAR-binding protein 1 or TRP 185.[10,27] Genetic evidence for interactions between loop-binding proteins and Tat also comes from studies of rodent cells where effects of Tat are minimal.[2] However, heterologous tethering of Tat via the coat protein, Rev or DNA-binding proteins (Gal 4, LexA) restores Tat transactivation in these cells. Further studies examining *in vivo* and *in vitro* binding interactions confirmed that a factor encoded by the human chromosome 12 could rescue this defect in rodent cells.[2] Thus, an intriguing notion is that loop-binding proteins or the complex of loop-binding proteins and Tat on TAR additionally require a protein from the human chromosome 12. The identification of this protein will be essential for the creation of small animal models of AIDS because even with recently isolated chemokine receptors, HIV will not replicate efficiently in rodent cells. Finally, loop-binding proteins isolated to date do not rescue this rodent defect.[10,27]

MECHANISM OF TAT TRANSACTIVATION

Tat is the only eukaryotic protein that activates transcription via RNA. However, there are several examples of such regulation in the prokaryotic world, e.g., the N protein of bacteriophage λ which antiterminates or elongates transcription via its binding to an RNA structure.[11] A similar mechanism pertains to Tat (Figure 5).[13,14] First, by nuclear run-on analyses, Tat does not increase rates of initiation but only elongation of transcription. Thus, the quantity of promoter proximal transcripts in the presence and absence of Tat is equivalent. However, the number of promoter distal transcripts is increased greatly by Tat. Since Tat via DNA can also increase rates of initiation, it has been postulated that Tat interacts with the same transcription factor differentially via DNA and RNA.[20] Tat also requires the C-terminal domain of RNA polymerase II for its effect and, as mentioned previously, can potentiate the kinase activity of TFIIH. Thus, a unifying hypothesis is that Tat interacts with components of the RNA polymerase II holoenyzme, which is assembled on the LTR in the absence of TAR and is repositioned by Tat on TAR, leading to the activation of TFIIH that phosphorylates the CTD. This kinase is CDK7 with which cyclin H and p36 form the CAK complex.[16] Upon its phosphorylation, the CTD loses associated initiation factors and assembles different proteins which lead to efficient elongation and cotranscriptional processing of messenger RNA by splicing and polyadenylation.[29]

(a)

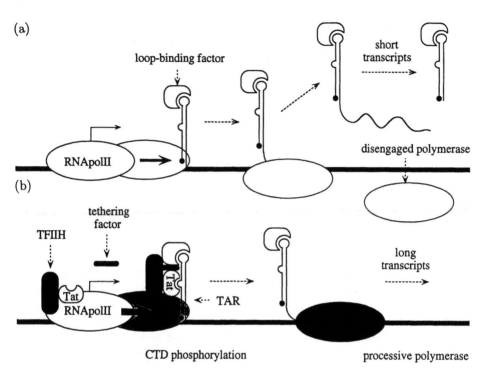

loop-binding factor

short transcripts

RNApolII

disengaged polymerase

(b)

tethering factor

TFIIH

Tat

Tat

RNApolII

<-- TAR

long transcripts

CTD phosphorylation   processive polymerase

FIGURE 5   Mechanism of transactivation by Tat. Although part of the RNA polymerase II holoenzyme, Tat does not initiate HIV transcription. Rather, proteins that interact with the LTR, such as TFIID, Initiator, Sp1, NFκB, etc., attract initiation complexes to the promoter. (a) In the absence of Tat, TAR is formed and attracts loop-binding proteins. Transcription complexes terminate prematurely, releasing short RNAs. (b) In the presence of Tat, TAR is able to direct TFIIH to the CTD and activate its kinase activity. Interactions between Tat and TAR result in the phosphorylation of the CTD, the transition from initiation to elongation of transcription, and the enzymatic conversion of RNA polymerase II to a processive form. Full-length, genomic viral transcripts are copied. The assembly of Tat-TAR complexes occurs cotranscriptionally and via nascent RNA.

Increasing evidence supports this model for the elongation of eukaryotic transcription. For example, when placed into promoters whose transcripts terminate prematurely, like the U2 SnRNA and c-myc promoters, Tat via TAR can antiterminate these transcription units.[26] Additionally, Tat does not increase rates of initiation from either of these promoters. In all these systems, the TATA box is absolutely essential for Tat transactivation *in vitro* and *in vivo*. This observation suggests that TBP and TAFs are required for initiating transcription complexes which are then modified by Tat as they pass through TAR. Thus, functions of initiation and elongation of transcription are separate and synergistic. In summary, Tat is the prototype for a cellular protein that affects the elongation of transcription. Therefore, Tat analogs must exist in eukaryotic systems, and it is possible that Tat will pave the way to their isolation and characterization.

IMPLICATIONS AND FUTURE DIRECTIONS

The study of transcriptional regulation of HIV is of great biological and clinical importance. For example, viruses that lack Tat are not pathogenic. Lesions in Tat are among the most deleterious for viral replication and gene expression.[13] Further mechanistic details of Tat transactivation should reveal potential new targets for interventions.

Of great interest to theoreticians is the interplay and kinetics between basal transcription factors that bind to DNA and Tat transactivation via RNA. As mentioned previously, they occur sequentially and potentiate each other. Mathematical models could predict what levels of activation by NFκB and NFAT are coupled to the synthesis of Tat and should reveal the delay in viral replication. Such modeling should present a two- or three-step model of viral transcriptional kinetics and define limiting steps along this pathway.

Future goals will also compare features of prokaryotic antitermination and eukaryotic elongation of transcription and define all the components required for initiation and elongation of HIV transcription in reconstituted systems. The isolation and characterization of proteins that interact with Tat and the CTD kinase, all of which are involved in Tat transactivation, will lead to structural analyses of interaction between Tat, TAR, and RNA polymerase II with the hope of finding small compounds that will inhibit their interactions. It is conceivable that by analyzing Tat, the mechanism(s) underlying the regulation of expression of cellular protooncogenes will also be revealed. Additionally, the copying of very long cellular genes will become better understood.

ACKNOWLEDGMENTS

We would like to acknowledge the expert secretarial assistance of Michael Armanini and helpful comments of members of our laboratory.

REFERENCES

1. Aboul-ela, F., J. Karn, and G. Varani. "Structure of HIV-1 TAR RNA in the Absence of Ligands Reveals a Novel Conformation of the Trinucleotide Bulge." *Nucleic Acids Res.* **24** (1996): 3974–3981.
2. Alonso, A., T. P. Cujec, and B. M. Peterlin. "Effects of Human Chromosome 12 on Interactions Between Tat and TAR of Human Immunodeficiency Virus Type 1." *J. Virol.* **68** (1994): 6505–6513.
3. Bayer, P., M. Kraft, A. Ejchart, M. Westendorp, R. Frank, and P. Rosch. "Structural Studies of HIV-1 Tat Protein." *J. Mol. Biol.* **247** (1995): 529–535.
4. Benko, D. M., S. Schwartz, G. N. Pavlakis, and B. K. Felber. "A Novel Human Immunodeficiency Virus Type 1 Protein, Tev, Shares Sequences with Tat, Env, and Rev Proteins." *J. Virol.* **64** (1990): 2505–2518.
5. Derse, D., M. Carvalho, R. Carroll, and B. M. Peterlin. "A Minimal Lentivirus Tat." *J Virol.* **65** (1991): 7012–7015.
6. Frankel, A. D. "Activation of HIV Transcription by Tat." *Curr. Opin. Genet. Dev.* **2** (1992): 293–298.
7. Frankel, A. D. "Peptide Models of the Tat-TAR Protein-RNA Interaction." *Protein Sci.* **1** (1992): 1539–1542.
8. Gait, M. J., and J. Karn. "Progress in Anti-HIV Structure-Based Drug Design." *Trends Biotechnol.* **13** (1995): 430–438.
9. Gait, M. J., and J. Karn. "RNA Recognition by the Human Immunodeficiency Virus Tat and Rev Proteins." *Trends Biochem. Sci.* **18** (1993): 255–259.
10. Gaynor, R. B. "Regulation of HIV-1 Gene Expression by the Transactivator Protein Tat." *Curr. Top. Microbiol. Immunol.* **193** (1995): 51–77.
11. Greenblatt, J., J. R. Nodwell, and S. W. Mason. "Transcriptional Antitermination." *Nature* **364** (1993): 401–406.
12. Ilyinskii P. O., M. D. Daniel, M. A. Simon, A. A. Lackner, and R. C. Desrosiers. "The Role of Upstream U3 Sequences in the Pathogenesis of Simian Immunodeficiency Virus-Induced AIDS in Rhesus Monkeys." *J. Virol.* **68** (1994): 5933–5944.
13. Jones K. A., and B. M. Peterlin. "Control of RNA Initiation and Elongation at the HIV-1 Promoter." *Ann. Rev. Biochem.* **63** (1994): 717–743.
14. Kao S., A. F. Calman, P. A. Luciw, and B. M. Peterlin. "Anti-Termination of Transcription Within the Long Terminal Repeat of HIV-1 by Tat Gene Product." *Nature* **330** (1987): 489–493.
15. Karn J., and M. A. Graeble. "New Insights Into the Mechanism of HIV-1 *Trans*-Activation." *Trends Genet.* **8** (1992): 36536–36538.
16. Morgan, D. O. "Principles of CDK Regulation." *Nature* **374** (1995): 131–134.
17. Mujeeb, A., K. Bishop, B. M. Peterlin, C. Turck, T. G. Parslow, and T. L. James. "NMR Structure of a Biologically Active Peptide Containing the

RNA-Binding Domain of Human Immunodeficiency Virus Type 1 Tat." *Proc. Natl. Acad. Sci. USA* **91** (1994): 8248–8252.

18. Parada, C. A., and R. G. Roeder. "Enhanced Processivity of RNA Polymerase II Triggered by Tat-Induced Phosphorylation of Its Carboxy-Terminal Domain." *Nature* **384** (1996): 375–378.

19. Pazin, M. J., P. L. Sheridan, K. Cannon, Z. Cao, J. G. Keck, J. T. Kadonaga, and K. A. Jones. "NF-Kappa B-Mediated Chromatin Reconfiguration and Transcriptional Activation of the HIV-1 Enhancer *In Vitro*." *Genes Dev.* **10** (1996): 37–49.

20. Pendergrast, P. S., and N. Hernandez. "RNA-Targeted Activators, But not DNA-Targeted Activators, Repress the Synthesis of Short Transcripts at the Human Immunodeficiency Virus Type 1 Long Terminal Repeat." *J. Virol.* **71** (1997): 910–917.

21. Sheldon, M., R. Ratnasabapathy, and N. Hernandez. "Characterization of the Inducer of Short Transcripts, a Human Immunodeficiency Virus Type 1 Transcriptional Element that Activates the Synthesis of Short RNAs." *Mol. Cell Biol.* **13** (1993): 1251–1263.

22. Sheridan, P. L., C. T. Sheline, K. Cannon, M. L. Voz, M. J. Pazin, J. T. Kadonaga, and K. A. Jones. "Activation of the HIV-1 Enhancer by the LEF-1 HMG Protein on Nucleosome-Assembled DNA *In Vitro*." *Genes Dev.* **9** (1995): 2090–2104.

23. Sullenger, B. A., H. F. Gallardo, G. E. Ungers, and E. Gilboa. "Overexpression of TAR Sequences Renders Cells Resistant to Human Immunodeficiency Virus Replication." *Cell* **63** (1990): 601–608.

24. Van Lint, C., S. Emiliani, M. Ott, and E. Verdin. "Transcriptional Activation and Chromatin Remodeling of the HIV-1 Promoter in Response to Histone Acetylation." *EMBO J.* **15** (1996): 1112–1120.

25. Verdin, E., and C. Van Lint. "Internal Transcriptional Regulatory Elements in HIV-1 and Other Retroviruses." *Cell. Mol. Biol.* **41** (1995): 365–369.

26. Wright, S., X. Lu, and B. M. Peterlin. "Human Immunodeficiency Virus Type 1 Tat Directs Transcription Through Attenuation Sites Within the Mouse c-myc Gene." *J. Mol. Biol.* **243** (1994): 568–573.

27. Wu-Baer, F., W. S. Lane, and R. B. Gaynor. "The Cellular Factor TRP-185 Regulates RNA Polymerase II Binding to HIV-1 TAR RNA." *EMBO J.* **14** (1995): 5995–6009.

28. Yang, X., C. H. Herrmann, and A. P. Rice. "The Human Immunodeficiency Virus Tat Proteins Specifically Associate with TAK *In Vivo* and Require the Carboxyl-Terminal Domain of RNA Polymerase II for Function." *J. Virol.* **70** (1996): 4576–4584.

29. Yuryev, A., M. Patturajan, Y. Litingtung, R. V. Joshi, C. Gentile, M. Gebara, and J. L. Corden. "The C-Terminal Domain of the Largest Subunit of RNA Polymerase II Interacts with a Novel Set of Serine/Arginine-Rich Proteins." *Proc. Natl. Acad. Sci. USA* **93** (1996): 6975–6980.

30. Zhou, Q., and P. A. Sharp. "Tat-SF1: Cofactor for Stimulation of Transcriptional Elongation by HIV-1 Tat." *Science* **274** (1996): 605–610.

Bryan R. Cullen
Howard Hughes Medical Institute and Department of Genetics, Duke University Medical
Center, Durham, NC 27710; E-mail: Culle002@mc.duke.edu

Role and Mechanism of Action of the HIV-1 Rev Regulatory Protein

Reverse transcription of the retroviral RNA genome results in the formation of a double-stranded DNA proviral intermediate that is then integrated into the genome of the host cell. The resultant provirus is flanked
by two long terminal repeats (LTRs). The 5′ LTR functions as the proviral transcriptional regulatory element while the 3′ LTR mediates efficient
polyadenylation of the resultant transcripts (Figure 1). The large majority of retroviruses contain this single LTR promoter element, so that the
primary transcript encoded by these retroviruses is identical to the viral
RNA genome. The presence of only one promoter element, when combined
with constraints on retroviral genome size imposed by the compact nature
of retroviral virions, has forced retroviruses to rely primarily on posttranscriptional mechanisms to regulate, and facilitate, the expression of the
various viral gene products. Of these mechanisms, the most important is
clearly regulated alternative splicing.

Viral Regulatory Structures and Their Degeneracies, edited by Gerald Myers.
SFI Studies in the Sciences of Complexity, Proc. Vol. XXVIII, Addison-Wesley, 1998   **85**

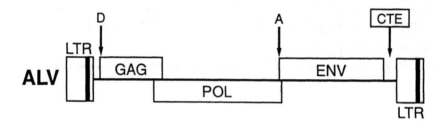

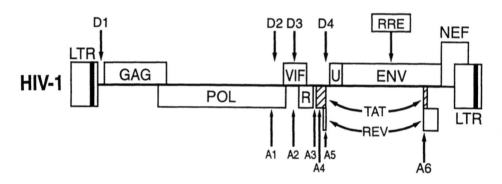

FIGURE 1   Genomic organization of ALV and HIV-1. Comparison of the genomic organization seen in ALV, a typical simple retrovirus, with the more complex HIV-1 genome. Viral genes are indicated, as are known major splice donor (D) and acceptor (A) sites. The localization of the Rev response element (RRE) and constitutive transport element (CTE) RNA targets are given. Abbreviations: LTR, long terminal repeat; R, Vpr; U, Vpu.

ALTERNATIVE SPLICING AND NUCLEAR RETENTION OF RETROVIRAL RNAs

Retroviruses do not encode any splicing factors, so that splicing of retroviral RNAs is controlled by the interaction of cellular splicing factors with *cis*-acting sequences in the viral transcript. In the case of the simple retrovirus avian leukemia virus (ALV), which encodes only a full length and a singly spliced transcript, an approximately 1:1 ratio between these mRNAs is maintained $1/M$ primarily as a result of a suboptimal splice acceptor site at the beginning of the viral *env* gene (Figure 1).[13] Mutations of this 3' splice site, which render its sequence closer to consensus, and, hence, improve splicing efficiency, result in an inhibition of viral replication and lead to the selection of viral revertants that, upon analysis, are found to again splice inefficiently.

The genome of human immunodeficiency virus type 1 (HIV-1) is significantly more complex than the genome of simple retroviruses such as ALV. In particular, while ALV encodes three gene products that are translated from only two mRNA species, HIV-1 encodes in excess of twenty mRNAs that permit the efficient translation of nine viral open reading frames (Figure 1). Despite this increased complexity, the basic mechanism remains the same, i.e., alternative splicing in HIV-1 is again regulated by the interplay of cellular transcription factors with suboptimal viral splice sites.[21]

The mRNAs encoded by HIV-1 can be divided into three classes: (1) the unspliced, ~9 kb transcript; (2) a set of five singly spliced transcripts of ~4 kb, derived by splicing from the major "D1" splice donor to one of the five splice acceptor sites (A1 to A5) located toward the center of the genome (Figure 1); and (3) a complex class of sixteen multiply spliced mRNAs, of ~2 kb in size, derived by exhaustive splicing of the 4 kb class. The 2 kb class encodes the viral regulatory proteins Tat, Rev, and Nef. The 4 kb class encodes the auxiliary proteins Vif, Vpr, and Vpu as well as the viral envelope protein, while the unspliced, genomic length mRNA not only encodes the Gag and Pol proteins but is also packaged into progeny virion particles.

An important distinction among these various viral mRNAs relates to whether they retain functional introns in their primary sequence. In particular, all members of the 2 kb class of viral mRNAs are fully spliced and retain no functional splice donor sequences and, hence, no definable introns. In contrast, both the unspliced 9 kb transcript and the five singly spliced viral RNAs, retain both functional splice donor sites and one or more identifiable introns. It is, of course, the complete removal of these introns that generates the fully spliced ~2 kb class of transcripts.

The reason that this difference is important is schematically illustrated in Figure 2, which shows that capped, polyadenylated mRNAs are normally efficiently and actively transported from the nucleus to the cytoplasm via channels in the nuclear membrane that are termed nuclear pores. However, if an RNA contains an identifiable intron, this induces an interaction with a subset of cellular splicing factors that have been termed commitment factors.[15] Although these factors have not been fully defined, they appear to include the U1 small nuclear ribonucleoprotein (snRNP) particle as well as members of the serine-arginine rich (SR) class of splicing factors. For most intron-containing RNAs, this commitment event simply represents the first step in the normal splicing pathway, leading inevitably to the appropriate assembly of further snRNPs and splicing factors on the RNA, to the RNA processing event itself and, once splicing is complete, to nuclear export (Figure 2). Alternatively, if the splice sites are recognized by the commitment machinery but are then found to be nonfunctional, the defective RNA can be degraded within the cell nucleus. Most importantly, however, recognition by commitment factors effectively blocks the nuclear export of the target RNA until and unless the intron(s) present in the RNA are fully removed.[3] The purpose of the cellular commitment machinery is, therefore, two-fold. First, these factors function in the identification

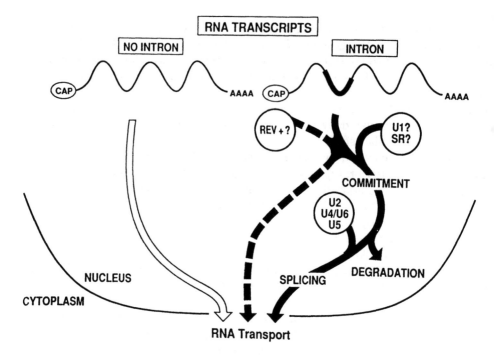

FIGURE 2   While fully spliced RNAs are readily exported from the nucleus, RNAs containing introns are retained in the nucleus by commitment factors, such as the U1 snRNP, until fully spliced or degraded. Rev induces the efficient nuclear export of RRE-containing target RNAs and thereby either prevents or reverses this nuclear retention activity.

and definition of introns and, thus, are normally critical for appropriate splicing to occur. Second, these factors effectively retain incompletely spliced cellular RNAs in the nucleus and, thus, prevent pre-mRNAs from encountering the cytoplasmic translational machinery of the cell. Clearly, translation of pre-mRNAs that retain introns within the intended protein coding sequence would generate defective proteins whose expression could be highly deleterious to the cell.

THE ROLE OF REV IN THE HIV-1 REPLICATION CYCLE

The existence of the splicing commitment pathway presents retroviruses in general, and HIV-1 in particular, with a serious conundrum. On the one hand, HIV-1 encodes a range of singly and multiply spliced mRNAs whose expression is critical to the viral life cycle. Therefore, the HIV-1 genome must contain splice sites that are recognizable by cellular splicing commitment factors. On the other hand, HIV-1 replication also requires the cytoplasmic expression of unspliced (9 kb) and incompletely spliced (4 kb) transcripts that retain these same splice sites. Therefore, the critical problem is: how do you express RNA transcripts that encode functional splice sites and yet avoid the nuclear retention of unspliced forms of these same transcripts by cellular commitment factors? As indicated in Figure 2, this problem is solved, in the case of HIV-1, by the Rev regulatory protein. As noted above, Rev is one of three viral proteins, along with Tat and Nef, that are encoded by the fully spliced ~2 kb class of viral mRNAs. Because these mRNAs are fully spliced, their nucleocytoplasmic transport and translation is constitutive in expressing cells. Once a significant level of Rev is expressed (this is dependent on the function of the Tat transcriptional transactivator encoded by HIV-1), it acts in the nucleus to induce the nucleocytoplasmic transport of incompletely spliced HIV-1 RNAs that contain the *cis*-acting Rev response element (RRE) RNA target for Rev.[4,18] The RRE is a highly ordered RNA stem-loop structure that is encoded within the body of the HIV-1 *env* gene (Figures 1 and 3). As such, it is present in all the incompletely spliced viral RNA transcripts whose expression is dependent on Rev.

The phenotype of the HIV-1 Rev protein is clearly revealed in Figure 4, which shows a northern analysis of cytoplasmic poly(A)[+] RNA. The cytoplasmic RNA probed in lane 2, derived from cells expressing a wild-type HIV-1 provirus, contains readily detectable levels of all three classes of viral mRNA. In contrast, the cytoplasmic RNA probed in lane 3, derived from a cell expressing an HIV-1 provirus bearing a defective *rev* gene, contains high levels of the fully spliced 2 kb class of viral transcripts but lacks any evidence for expression of the incompletely spliced 9 kb and 4 kb viral mRNA species. Importantly, published analyses of nuclear viral mRNA expression patterns[4,18] clearly demonstrate substantial levels of expression of these 9 kb and 4 kb viral RNAs in the nuclei of cells both in the presence and absence of Rev. There is, therefore, no evidence based on these data to suggest that Rev interferes with splicing directly. Instead, these observations clearly suggest that Rev functions as an RNA sequence-specific nuclear export factor.

More recently, an elegant analysis of Rev function using microinjected Xenopus oocytes has unequivocally demonstrated that Rev can directly induce the nuclear export of target RNAs from the nucleus.[6] Of interest, Rev was able to induce the efficient nuclear export of not only RNAs retained by splicing commitment factors, but also RNAs that are inefficiently exported for other reasons, such as the absence of an appropriate 5′ cap structure. Overall, these data have clearly fully validated the earlier proposal that Rev is a sequence-specific nuclear export factor.

Before moving on to a discussion of the mechanism of action of Rev, I would briefly like to discuss the question of how simple retroviruses, such as ALV, promote the nuclear export of their genome length RNA transcript. In principle, in this simpler system one could envision several possible mechanisms. For example, the single-splice donor in ALV could form part of an extended RNA secondary

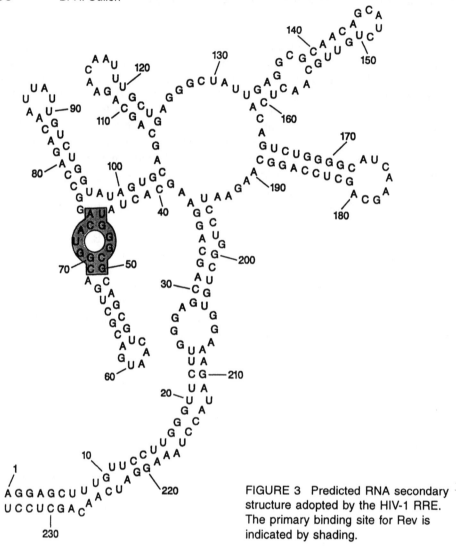

FIGURE 3 Predicted RNA secondary structure adopted by the HIV-1 RRE. The primary binding site for Rev is indicated by shading.

that could randomly fold into two distinct conformations. In one case, the splice donor could be sequestered in an RNA stem, a location known to block recognition by splicing factors. In the second conformation, it could be fully exposed and, hence, efficiently recognized by splicing commitment factors. Although this model remains possible for some simple retroviruses, it is clearly not valid for avian retroviruses such as ALV or for type D retroviruses such as Mason-Pfizer monkey virus (MPMV). Instead, these viruses contain a *cis*-acting structured RNA target sequence, termed the constitutive transport element (CTE), that functions like the HIV-1 RRE to promote cytoplasmic expression of unspliced RNA transcripts (see Felber, this volume).[2,20] However, neither ALV nor MPMV encode a protein able to interact with the CTE, and it therefore appears that the CTE is a target site for a cellular Rev-like RNA binding protein. In both ALV and in MPMV, the CTE is located between the *env* gene and the LTR (Figure 1). Deletion of the CTE, as one might predict, blocks the nucleocytoplasmic transport of the unspliced viral genomic RNA but does not prevent expression of the fully spliced *env* transcript.

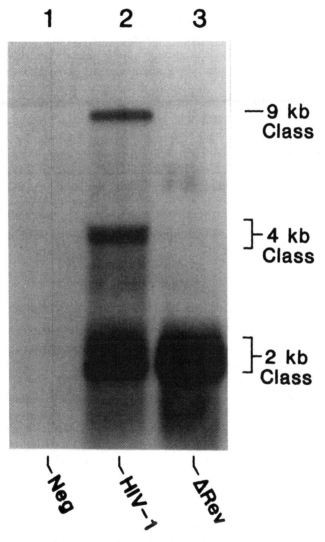

FIGURE 4   Northern analysis of cytoplasmic poly(A)$^+$ RNA derived from cells expressing a wild-type (lane 2) or Rev-defective (lane 3) HIV-1 provirus. Cytoplasmic expression of the incompletely spliced 4 kb and 9 kb classes of HIV-1 RNA is dependent on Rev function. Lane 1 contains RNA derived from uninfected cells.

Although the CTE and the RRE therefore appear functionally analogous, an impression reinforced by the finding that the CTE can functionally substitute for the RRE in mediating HIV-1 structural gene expression, there is a significant practical difference between these two posttranscriptional regulatory pathways. In particular, the fact that HIV-1 encodes its own Rev protein, while ALV and MPMV rely on a presumably similar protein encoded by the cell, markedly affects the temporal order of viral protein expression. Thus, the HIV-1 replication cycle can be divided into an early, Rev-independent phase and a late, Rev-dependent phase.[14,18] This pattern of early expression of viral regulatory proteins followed by late expression of structural proteins is widely seen in virology and is indeed typical of DNA tumor viruses, including papillomaviruses (see McBride et al. and Baker, this volume). It is likely that this delay in the expression of the viral structural proteins, which are the primary targets for immune surveillance by the host, confers a significant replicative advantage on the viruses that have developed this temporal regulation. In contrast, the CTE, because it acts via a preexisting cellular cofactor, does not support such an early/late separation. While this is perhaps not a problem for simple retroviruses such as ALV, it is conceivable that this lack could represent a significant disadvantage for complex retroviruses. It is, therefore, possible that the CTE seen in simple retroviruses such as ALV and MPMV represents an evolutionary precursor to the more autonomous Rev:RRE regulatory loop seen in all lentiviruses, including HIV-1, as well as the very similar Rex:Rex Response Element axis seen in the various T-cell leukemia viruses.

DOMAIN ORGANIZATION OF HIV-1 REV

The elucidation of the functional domain organization of Rev (Figure 5) has provided considerable insight into the mechanism of action of this novel regulatory protein.[16,17] A highly basic, arginine-rich sequence located toward the amino-terminus of Rev constitutes a nuclear localization signal (NLS) and also forms the sequence-specific RNA binding domain of Rev. Extensive *in vitro* and *in vivo* analyses of the interaction of Rev with the RRE have demonstrated that Rev first binds to a discrete, highly structured RNA bulge within the RRE (Figure 3).[11,24] Subsequently, additional Rev molecules are recruited to the RRE in a multimerization process that is mediated by both protein:protein and protein:RNA interactions.[11,17] Protein sequences critical to this essential multimerization event have been localized to either side of the Rev RNA binding domain (Figure 5). While these additional bound Rev molecules appear to occupy discrete sites on the RRE, as determined by RNAase footprinting, no specific RNA target sequence has been identified. These secondary RNA binding events therefore appear to be distinct from Rev binding to its primary target site, which is clearly highly sequence-specific.[11,24]

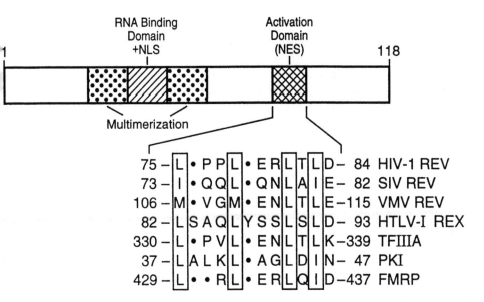

FIGURE 5 Functional domain organization of the HIV-1 Rev protein. Several sequences derived from other viral or cellular proteins that have been shown to be able to functionally substitute for the Rev activation domain/nuclear export signal (NES) are shown. Critical large hydrophobic residues in the NES are boxed. NLS, nuclear localization signal.

Because Rev multimerization on the RRE is critical for Rev function, it is apparent that the active form of Rev, at least as far as nuclear RNA export is concerned, is in the form of a ribonucleoprotein particle.

Further mutational analysis of Rev has shown that a small leucine-rich sequence, termed the activation domain, is also critical for Rev function but is dispensable for both RNA binding and protein multimerization (Figure 5).[16,17] Analysis of Rev proteins derived from other lentiviruses, including simian immunodeficiency virus (SIV) and sheep visna maedi virus (VMV), has identified functionally equivalent domains within these distantly related proteins and shown that chimeric proteins containing the Rev RNA binding/multimerization domain linked to these heterologous activation domains are fully active (Figure 5). Sequences able to effectively substitute for the Rev activation domain have also been identified in the functionally similar Rex protein, encoded by human T-cell leukemia virus type I (HTLV-I) as well as in several cellular proteins, including amphibian transcription factor IIIA (TFIIIA), human protein kinase inhibitor α (PKI) and human fragile X-mental retardation protein (FMRP).[7,12]

Mutational analysis of the ~10 amino acid Rev activation domain soon demonstrates that the four leucine residues present in this sequence are critical for its activity, although substitution with other large hydrophobic residues such as isoleucine, valine or methionine has been tolerated at some positions. This importance is also reflected in the high level of conservation of these residues in functionally equivalent sequences of different origin (Figure 5).

A particularly informative observation resulting from this mutational analysis is the demonstration that Rev proteins bearing a functional RNA binding/multi-merization sequence but lacking an intact activation domain displayed a potent dominant negative phenotype, i.e., they could effectively block the activity of wild-type Rev when present in *trans*.[16] This finding has obvious potential application in the gene therapy of HIV-1 induced disease and is being actively pursued by several clinically oriented groups. In addition, this finding strongly suggests that the activation domain is a critical cofactor binding domain. It is believed that mutant Rev proteins lacking a functional activation domain participate, with wild-type Rev, in multimerization on the RRE RNA target and then interfere with the biological activity of the wild-type protein by inhibiting the recruitment of relevant cellular cofactors to the resultant ribonucleoprotein complex.

THE REV ACTIVATION DOMAIN IS A NUCLEAR EXPORT SIGNAL

If Rev indeed functions as a sequence-specific nuclear RNA export factor, then one might expect that Rev would be exported to the cytoplasm along with its RRE-containing RNA target (Figure 2). Presumably, Rev would then release the RNA into the cytoplasm and return to the nucleus upon recognition of its exposed NLS sequence (Figure 5). In fact, Meyer and Malim[19] have demonstrated that Rev rapidly shuttles back and forth between the nucleus and the cytoplasm. Interestingly, Rev mutants lacking a functional activation domain fail to shuttle, thus implying that this short sequence actually targets Rev to the cytoplasm. This hypothesis has now been confirmed by several groups by nuclear microinjection of various carrier proteins linked to the Rev activation domain.[5,25] As shown in Figure 6(a), such fusion proteins (in this case glutathione-S-transferase [GST] linked to the Rev activation domain) are rapidly exported to the cell cytoplasm. In contrast, a very similar fusion protein, consisting of GST linked to the defective M10 mutant form of the Rev activation domain, is retained in the nucleus (Figure 6(c)). Therefore, it is apparent that the Rev activation domain forms an autonomous protein nuclear export signal (NES). Subsequent work has demonstrated that other Rev activation domains, such as those seen in SIV and VMV Rev, as well as the activation domain observed in the functionally analogous HTLV-I Rex protein, also function as NESs. Similarly, the leucine-rich activation domain found in the cellular protein TFIIIA also functions as an efficient nuclear export signal[7] (Figure 6(e)), while a mutant form of this TFIIIA sequence fails to mediate nuclear export of the

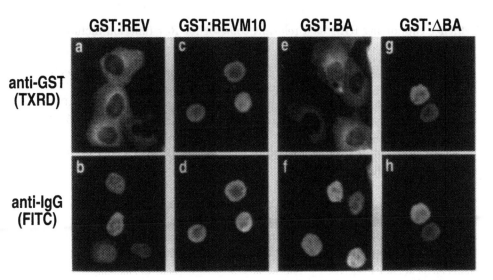

FIGURE 6   Microinjection analysis of NES activity. Fusion proteins were prepared consisting of glutathione-S-transferase (GST) fused to residues 67 to 116 of wild-type Rev or the Rev M10 mutant (M10 contains aspartic acid-leucine in place of residues 78 and 79—leucine/glutamic acid—in the Rev activation domain) or residues 321 to 339 of wild-type (GST:BA) or mutant (alanine in place of leucine 336—GST: Δ BA) forms of Bufo americanus TFIIIA. These were then mixed with rabbit IgG and injected into the nucleus of HeLa cells. After 30 minutes of incubation at 37°C, the subcellular localization of the GST fusion proteins, and of the IgG internal control, were determined by immunofluorescence. Fusion proteins containing the wild-type Rev or TFIIIA sequences were efficiently exported to the cytoplasm (panels (a) and (e)) while the mutants were retained in the nucleus (panels (c) and (g)). Reproduced from Fridell et al.[7] Permission granted by *Proc. Natl. Acad. Sci. USA* **93** (1996): 2936–2940.

linked GST protein (Figure 6(g)). In addition, the functionally equivalent sequences identified in the human PKI and FMRP proteins have also been shown to function as NESs.[25] Not surprisingly, mutational analysis indicates that there is a complete correlation between the ability of any of these sequences to function as an NES and their ability to function as a Rev activation domain.

THE REV ACTIVATION DOMAIN BINDS TO NUCLEOPORINS

The obvious next question is the nature of the cellular target for the Rev activation domain/NES. This problem has been resolved essentially simultaneously by three groups, who each independently demonstrated that the HIV-1 Rev NES is able to specifically interact with a nucleoporin-like cellular factor termed the Rev interacting protein (RIP) or the   Rev activation domain binding (RAB) protein.[1,9,23]

Evidence supporting RAB as a relevant cofactor for Rev function includes the following

1. Analysis of a wide range of Rev NES mutants demonstrated a complete correlation between RAB binding and activation domain function.
2. RAB was able to bind Rev specifically when Rev was bound to the RRE.
3. RAB bound to other retroviral and cellular NES sequences (Figure 5), but not to defective mutants.
4. Overexpression of RAB modestly but significantly enhanced Rev function in transfected cells.

The human RAB protein is a ~58 kDa protein localized to the nucleus and, potentially, the nuclear membrane of expressing cells. Importantly, RAB has several features typical of cellular nucleoporins, including numerous repeats of the highly characteristic dipeptide sequence phenylalanine-glycine (FG). Mutational analysis has demonstrated that the FG repeat elements present in RAB form a critical part of the binding site for the Rev NES sequence.

Nuclear pores are large, highly ordered structures, containing 100 or more distinct cellular proteins, termed nucleoporins, that control the entry/exit of molecules from the nucleus of the cell.[10] A general characteristic of nucleoporins is the presence of multiple copies of the FG motif, frequently in the context of the longer repeat elements FXFG or GLFG. Recently it has been demonstrated that many, but not all, cellular nucleoporins are also able to specifically bind to functional NES sequences, including the HIV-1 Rev NES, but not to nonfunctional NES mutants.[8,22] The affinity of the Rev NES for these cellular nucleoporins appears comparable to the Rev:Rab interaction and is again mediated by sequences containing multiple FG repeats. It is, therefore, apparent that the molecular targets for the Rev NES include constituents of the nuclear pore, the very structure that is known to regulate nuclear RNA and protein export. In addition, the fact that several cellular proteins have now been shown to contain leucine-rich NES sequences, that also interact with RAB and cellular nucleoporins,[8] clearly demonstrates that the nuclear export pathway utilized by Rev is, in fact, an important mechanism for the appropriate subcellular localization of cellular proteins and, potentially, RNAs.

A MODEL FOR THE MECHANISM OF ACTION OF REV

The mechanisms regulating the nuclear import of cellular or viral proteins are fairly well understood and provide a relevant model for understanding the nuclear export pathway utilized by Rev-like NES sequences. It is now well established that basic NLS sequences first interact with a cytoplasmic receptor called karyopherin α or importin α.[10] This heterodimer binds a second cellular factor, termed karyopherin β or

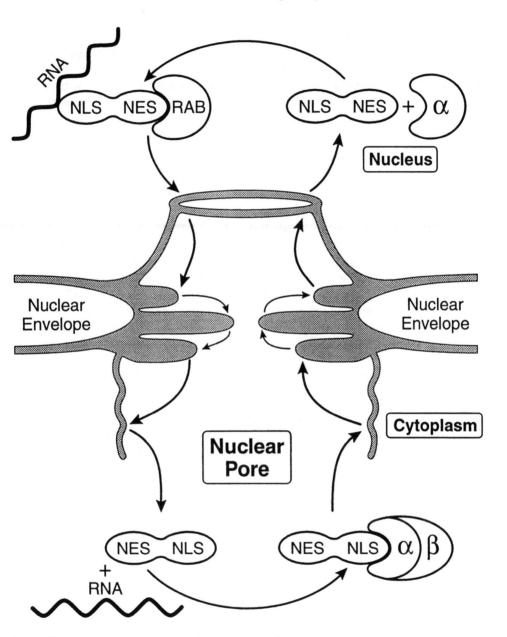

FIGURE 7   Possible mechanism of action of the Rev NES sequence. See text for detailed discussion.

importin β (Figure 7). Karyopherin then conveys the resultant heterotrimer to the nuclear pore, where the β subunit docks by binding to cellular nucleoporins. Subsequently, the NLS receptor complex is hypothesized to move through the nuclear pore due to the sequential interaction of the karyopherin β subunit with nucleoporins located at different sites in the pore. This movement requires the activity of the cellular GTPase Ran and is dependent on energy derived by hydrolysis of GTP. Once the heterotrimer reaches the inner face of the nuclear membrane, a direct interaction been Ran:GTP and β-karyopherin is believed to result in the release of the NLS protein, and karyopherin α, into the nucleoplasm (Figure 7).

Based on the observations described above, one can propose a somewhat similar mechanism of action for the Rev-like NES sequences.[8,10] The first step in this pathway would be the assembly of Rev onto the RRE RNA followed by the recruitment of the primarily nucleoplasmic Rab cofactor (Figure 7). Rab is hypothesized to facilitate the movement of the Rev ribonucleoprotein complex to the nuclear pore, where the Rev NES would then dock via a direct interaction with a nucleoporin. Subsequently, the NES-containing complex would again move directionally through the nuclear pore as the result of the sequential interaction of Rev with specific nucleoporins. The export of NES-containing proteins is known to require energy,[19] but it remains unclear whether this involves one or more specific Ran-like cofactors. Finally, once the NES complex reaches the cytoplasm it is disassembled, thus exposing the Rev NLS and inducing its nuclear import (Figure 7). By this means, Rev would continuously shuttle between nucleus and cytoplasm.

While the model for Rev export presented in Figure 7 is very attractive, many questions remain unanswered. For example, how can Rev enter the nucleus bearing a functional NES? Is the NES somehow occluded during nuclear import? What is the energy source for export and how is directionality assured? How is Rev released from the nuclear pore, and once in the cytoplasm, from the RNA target? Clearly, these and many other questions remain to be fully addressed. Nevertheless, it is apparent that work on the HIV-1 Rev regulatory protein has provided a remarkable harvest of insights into a cellular mechanism, i.e., regulated nuclear export of proteins and RNAs, that until recently had remained totally obscure. It is therefore apparent that the study of viruses continues to provide important and surprising insights into the biology of the eukaryotic cell.

REFERENCES

1. Bogerd, H. P., R. A. Fridell, S. Madore, and B. R. Cullen. "Identification of a Novel Cellular Cofactor for the Rev/Rex Class of Retroviral Regulatory Proteins." *Cell* **82** (1995): 485–494.
2. Bray, M., S. Prasad, J. W. Dubay, E. Hunter, K.-T. Jeang, D. Rekosh, and M.-L. Hammarskjöld. "A Small Element from the Mason-Pfizer Monkey Virus Genome Makes Human Immunodeficiency Virus Type 1 Expression and Replication Rev-Independent." *Proc. Natl. Acad. Sci. USA* **91** (1994): 1256–1260.
3. Chang, D. D., and P. A. Sharp. "Regulation by HIV Rev Depends Upon Recognition of Splice Sites." *Cell* **59** (1989): 789–795.
4. Felber, B. K., M. Hadzopoulou-Cladaras, C. Cladaras, T. Copeland, and G. N. Pavlakis. "Rev Protein of Human Immunodeficiency Virus Type 1 Affects the Stability and Transport of the Viral mRNA." *Proc. Natl. Acad. Sci. USA* **86** (1989): 1495–1499.
5. Fischer, U., J. Huber, W. C. Boelens, I. W. Mattaj, and R. Lührmann. "The HIV-1 Rev Activation Domain is a Nuclear Export Signal that Accesses an Export Pathway Used by Specific Cellular RNAs." *Cell* **82** (1995): 475–483.
6. Fischer, U., S. Meyer, M. Teufel, C. Heckel, R. Lührmann, and G. Rautmann. "Evidence that HIV-1 Rev Directly Promotes the Nuclear Export of Unspliced RNA." *EMBO J.* **13** (1994): 4105–4112.
7. Fridell, R. A., U. Fischer, R. Lührmann, B. E. Meyer, J. L. Meinkoth, M. H. Malim, and B. R. Cullen. "Amphibian Transcription Factor IIIA Proteins Contain a Sequence Element Functionally Equivalent to the Nuclear Export Signal of Human Immunodeficiency Virus Type 1 Rev." *Proc. Natl. Acad. Sci. USA* **93** (1996): 2936–2940.
8. Fritz, C. C., and M. R. Green. "HIV Rev Uses a Conserved Cellular Protein Export Pathway for the Nucleocytoplasmic Transport of Viral RNAs." *Current Biol.* **6** (1996): 848–854.
9. Fritz, C. C., M. L. Zapp, and M. R. Green. "A Human Nucleoporin-Like Protein that Specifically Interacts with HIV Rev." *Nature* **376** (1995): 530–533.
10. Görlich, D., and I. W. Mattaj. "Nucleocytoplasmic Transport." *Science* **271** (1996): 1513–1518.
11. Heaphy, S., J. T. Finch, M. J. Gait, J. Karn, and M. Singh. "Human Immunodeficiency Virus Type 1 Regulator of Virion Expression, Rev, Forms Nucleoprotein Filaments After Binding to a Purine-Rich 'Bubble' Located Within the Rev-Responsive Region of Viral mRNAs." *Proc. Natl. Acad. Sci. USA* **88** (1991): 7366–7370.
12. Hope, T. J., B. L. Bond, D. McDonald, N. P. Klein, and T. G. Parslow. "Effector Domains of Human Immunodeficiency Virus Type 1 Rev and Human T-Cell Leukemia Virus Type I Rex Are Functionally Interchangeable and Share an Essential Peptide Motif." *J. Virol.* **65** (1991): 6001–6007.

13. Katz, R. A., and A. M. Skalka. "Control of Retroviral RNA Splicing Through Maintenance of Suboptimal Processing Signals." *Mol. Cell. Biol.* **10** (1990): 696–704.

14. Kim, S., R. Byrn, J. Groopman, and D. Baltimore. "Temporal Aspects of DNA and RNA Synthesis During Human Immunodeficiency Virus Infection: Evidence for Differential Gene Expression." *J. Virol.* **63** (1989): 3708–3713.

15. Legrain, P., and M. Rosbash. "Some *Cis*- and *Trans*-Acting Mutants for Splicing Target Pre-mRNA to the Cytoplasm." *Cell* **57** (1989): 573–583.

16. Malim, M. H., S. Böhnlein, J. Hauber, and B. R. Cullen. "Functional Dissection of the HIV-1 Rev *Trans*-Activator—Derivation of a *Trans*-Dominant Repressor of Rev Function." *Cell* **58** (1989): 205–214.

17. Malim, M. H., and B. R. Cullen. "HIV-1 Structural Gene Expression Requires the Binding of Multiple Rev Monomers to the Viral RRE: Implications for HIV-1 Latency." *Cell* **65** (1991): 241–248.

18. Malim, M. H., J. Hauber, S.-Y. Le, J. V. Maizel, and B. R. Cullen. "The HIV-1 Rev Trans-Activator Acts Through a Structured Target Sequence to Activate Nuclear Export of Unspliced Viral mRNA." *Nature* **338** (1989): 254–257.

19. Meyer, B. E., and M. H. Malim. "The HIV-1 Rev Trans-Activator Shuttles Between the Nucleus and the Cytoplasm." *Genes Dev.* **8** (1994): 1538–1547.

20. Ogert, R. A., L. H. Lee, and K. L. Beemon. "Avian Retroviral RNA Element Promotes Unspliced RNA Accumulation in the Cytoplasm." *J. Virol.* **70** (1996): 3834–3843.

21. Staffa, A., and A. Cochrane. "The Tat/Rev Intron of Human Immunodeficiency Virus Type 1 is Inefficiently Spliced Because of Suboptimal Signals in the 3' Splice Site." *J. Virol.* **68** (1994): 3071–3079.

22. Stutz, F., E. Izaurralde, I. W. Mattaj, and M. Rosbash. "A Role for Nucleoporin FG Repeat Domains in Export of Human Immunodeficiency Virus Type 1 Rev Protein and RNA from the Nucleus." *Mol. Cell. Biol.* **16** (1996): 7144–7150.

23. Stutz, F., M. Neville, and M. Rosbash. "Identification of a Novel Nuclear Pore-Associated Protein as a Functional Target of the HIV-1 Rev Protein in Yeast." *Cell* **82** (1995): 495–506.

24. Tiley, L. S., M. H. Malim, H. K. Tewary, P. G. Stockley, and B. R. Cullen. "Identification of a High-Affinity RNA-Binding Site for the Human Immunodeficiency Virus Type 1 Rev Protein." *Proc. Natl. Acad. Sci. USA* **89** (1992): 758–762.

25. Wen, W., J. L. Meinkoth, R. Y. Tsien, and S. S. Taylor. "Identification of a Signal for Rapid Export of Proteins from the Nucleus." *Cell* **82** (1995): 463–473.

Barbara K. Felber
ABL-Basic Research Program, Bldg. 535, Rm. 110, NCI-FCRDC, P.O. Box B, Frederick, MD
21702–1201; E-mail: felber@ncifcrf.gov

Posttranscriptional Control: A General and Important Regulatory Feature of HIV-1 and Other Retroviruses

Posttranscriptional regulation is a key event in the expression of many cellular mRNAs, and it is also an essential mechanism for the expression of several retroviruses and DNA viruses (Table 1). Studies of the human immunodeficiency virus type 1 (HIV-1) and the human T-cell leukemia virus type I (HTLV-I) showed that these retroviruses depend on virally regulated posttranscriptional regulation for their expression (see Cullen, this volume). Recent findings demonstrate that simple retroviruses such as the simian type D retroviruses (SRV) and avian leukosis-sarcoma viruses (ALV) also regulate their expression at the posttranscriptional level. And further, it has been shown that several intracisternal A-particle retroelements contain a functional posttranscriptional element related to that of the type D retroviruses. Retroviral posttranscriptional regulation is mediated by the interaction of a distinct *cis*-acting viral RNA element with viral proteins or with cellular factors that are of ubiquitous or species-specific nature. Although the detailed mechanism of this regulation is different among the individual retroviruses, in all cases it is essential for at least the transport and expression of the full-length mRNA. Recent findings have shown that expression of the papillomavirus structural L1 protein is also regulated at the posttranscriptional level (Table 1, see also Baker, this volume). It remains to be seen whether the type B and the type C oncoretroviruses, the spumaviruses or the polyomaviruses undergo similar regulation.

In this chapter, we review the posttranscriptional mechanisms used by the different retroviruses. An important conclusion is that several retroviruses depend on posttranscriptional regulation for their expression. As a general rule, this regulation is essential for the expression of the *gag/pol* genes, which are encoded by the full-length mRNA. These findings have led to a revision of the previous model suggesting that the expression of retroviral structural proteins depends only on the presence of suboptimal splice sites. It has been previously hypothesized that suboptimal splice sites, responsible for slow splicing of the genomic RNA, are sufficient to allow the nucleocytoplasmic transport of the unspliced transcript.[12] Evidence from several retrovirus systems indicates that this transport is an active process mediated through elaborate posttranscriptional control mechanisms. Elucidation of the mechanisms of function of the different viral regulatory systems will be important for the understanding of virus propagation and of the mechanisms of nucleocytoplasmic transport and expression of cellular mRNAs.

THE ROLE OF THE CONSTITUTIVE TRANSPORT ELEMENT (CTE) IN MEDIATING POSTTRANSCRIPTIONAL REGULATION OF THE TYPE D SIMIAN RETROVIRUSES

The type D simian retroviruses (SRV) is a family of retroviruses consisting of simian retrovirus type 1 (SRV-1), type 2 (SRV-2), and type 3 (SRV-3) or Mason-Pfizer monkey virus (MPMV). These viruses belong to the simple retroviruses encoding *gag*, *pol*, and *env* and no other viral proteins (Figure 1, Table 1). Gag is produced as a polyprotein that upon proteolysis forms the capsid of the virion. The *pol* gene encodes the enzymes Protease, Reverse Transcriptase, and Integrase. Env forms the layer of the surface glycoprotein. Typical to this class of retroviruses, these genes are encoded by two transcripts. The *gag/pol* genes are encoded by the unspliced full-length mRNA, whereas the *env* gene is encoded by the singly spliced mRNA. Therefore, the unspliced retroviral transcript serves as genomic RNA that is packaged into the progeny virions, as mRNA for the *gag/pol* genes, as well as precursor for the singly spliced *env* mRNA. Recent studies have demonstrated that the transport of this RNA is mediated via a small *cis*-acting RNA element, termed constitutive transport element (CTE), that interacts with putative cellular factors. The CTE is located between the terminator of *env* and the 3' long terminal repeat (LTR) and spans approximately 170 nucleotides in length.[4,21,26] These findings demonstrate that the type D simian retroviruses depend on posttranscriptional regulation for their expression. Although the CTE is also present at the 3′ end of both mRNAs, the singly spliced *env* mRNA can be transported to the cytoplasm

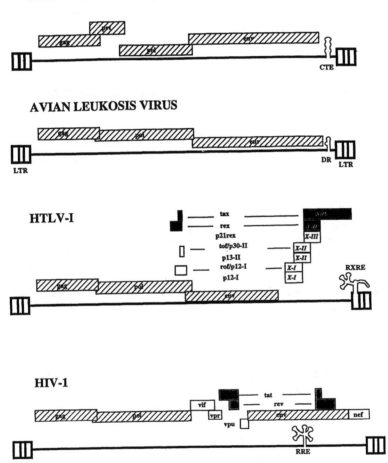

FIGURE 1 Genomes of simian type D retroviruses (SRV), avian leukosis-sarcoma virus (ALV), human T-cell leukemia virus type I (HTLV-I) and human immunodeficiency virus type 1 (HIV-1) are shown. All retroviruses share the genes encoding the structural proteins Gag/Pol and Env. HTLV-I and HIV-1 encode also regulatory proteins (Tax, Rex and Tat, Rev, respectively) and several accessory proteins. The posttranscriptional control elements, CTE in SRV and DR in ALV, are located beteeen *env* and the 3′ LTR. In contrast the RXRE of HTLV-I is located in the 3′ LTR, whereas the RRE of HIV-1 is located within the *env* gene.

TABLE 1

RETROVIRUSES		Posttranscriptional Control	Mechanism
Lentiviruses	HIV-1, HIV-2, SIV, FIV, BIV, VISNA, EIAV, CAEV	YES	RRE/Rev
Oncoretroviruses			
Avian leukosis-sarcoma	ALV	YES	RNA elements/avian cell specific factor(s)
Type B	MMTV	?	
Type C	MuLV	?	
Type D	SRV-1, SRV-2, MPMV	YES	CTE/cellular factor(s)
HTLV/BLV	HTLV-I, HTLV-II, BLV	YES	RxRE/Rex
Spumaviruses	HFV	?	
Retroelements	IAP	YES	CTE/cellular factor(s)
PAPOVAVIRUSES			
Papillomaviruses	HPV, BPV	YES	RNA elements?/cell specific factor(s)
Polyomaviruses		?	

and expressed in the absence of the CTE, and, therefore, Env is produced independently of posttranscriptional regulation.[8] In contrast, the CTE is essential for the production of Gag/Pol proteins and, hence, for the production of infectious type D retrovirus.[8,20] This observation points to the fact that elements within the 5' portion of the *gag/pol* mRNA are responsible for the nuclear retention of the unspliced RNA. The role of CTE in the type D retroviruses is, therefore, solely to ensure the transport of the unspliced transcript from the nucleus to the cytoplasm.

SRV-1 and MPMV are more closely related than SRV-2. Comparison of the *gag/pol* and *env* genes between SRV-1 and MPMV shows similarity of approximately 90–97%, whereas there is about a 72–80% similarity between SRV-1 to SRV-2. Among the CTEs of these three viruses there is a high degree of sequence similarity (88–92%). The CTEs were shown to fold into an extended RNA stem-loop structure (Figure 2).[8,21] An important characteristic of these RNA elements is the presence of two highly conserved internal loop regions, which are separated by a stem structure of equal length, and a less-conserved hairpin loop. Comparison of the

SRV-1 CTE to those of MPMV and SRV-2 reveals further that nucleotide changes in the MPMV and SRV-2 elements are accompanied by compensatory changes in the double-stranded regions, resulting in conservation of the structure. The definition of the molecular and structural requirements for function of this RNA element showed that both sequence (internal loops) and secondary structure (including the length of the stem structure separating the loop regions) are important for appropriate function.[8,21] Since the type D retrovirus CTE and a functional CTE-like element identified in an intracisternal A-particle retroelement[20] (see below) share only the sequence of these internal loop regions and the overall structure of the element, the loop regions are thought to represent the interaction sites with the putative cellular factors. These internal loops are formed by two imperfect direct repeats and as a result, it is hypothesized that the factors interacting with the two loops may form a dimer. This would be feasible because of the close proximity of the loops, which are separated by a stem structure having three bulges. CTE-mediated posttranscriptional regulation occurs in several cell types and species which are not the target for infection by the type D retroviruses.[24] Since the machinery for CTE-mediated expression is present ubiquitously in the mammalian cells, this finding further suggests that CTE-like elements exist in some cellular RNAs.

In summary, the genomic organization of the simian type D retroviruses does not have the complexity of lentiviruses and HTLV oncoretroviruses (Figure 1). Nevertheless, the production of type D virions depends on complex posttranscriptional regulation and on the *cis*-acting viral RNA element CTE interacting

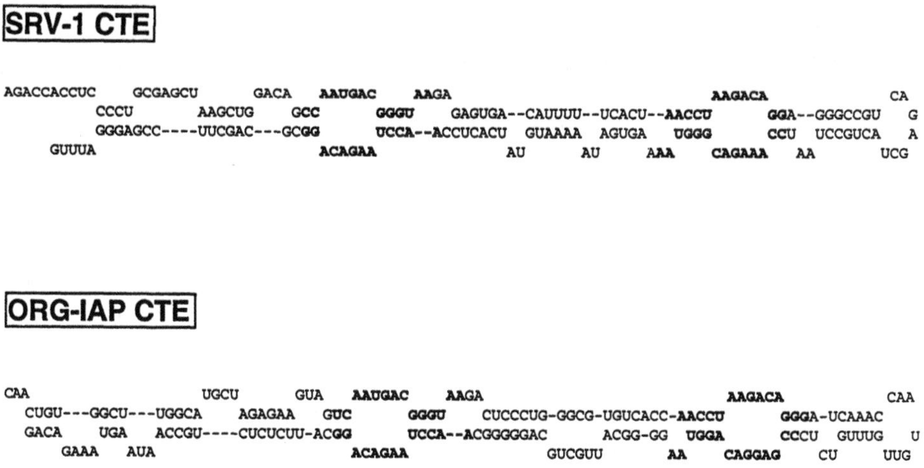

SRV-1 CTE

```
AGACCACCUC    GCGAGCU    GACA   AAUGAC   AAGA                                        AAGACA           CA
         CCCU        AAGCUG   GCC    GGGU   GAGUGA--CAUUUU--UCACU--AACCU   GGA--GGGCCGU   G
         GGGAGCC----UUCGAC---GCGG   UCCA--ACCUCACU  GUAAAA  AGUGA   UGGG   CCU  UCCGUCA   A
    GUUUA                          ACAGAA           AU      AU      AAA    CAGAAA   AA        UCG
```

ORG-IAP CTE

```
CAA                      UGCU     GUA   AAUGAC   AAGA                                        AAGACA           CAA
   CUGU---GGCU---UGGCA    AGAGAA   GUC    GGGU   CUCCCUG-GGCG-UGUCACC-AACCU   GGGA-UCAAAC
   GACA    UGA   ACCGU----CUCUCUU-ACGG   UCCA--ACGGGGGAC         ACGG-GG  UGGA   CCCU  GUUUG   U
     GAAA    AUA                         ACAGAA            GUCGUU    AA     CAGGAG   CU      UUG
```

FIGURE 2 The structures of the SRV-1 CTE and the ORG-CTE are shown. These CTEs share the sequence of the internal loop regions (in bold) and the overall secondary structures.

with the cellular posttranscriptional machinery. The identification of cellular RNA homologs of the CTE and the cellular machinery mediating CTE function will further increase our understanding of the mechanisms involved in viral and cellular posttranscriptional regulation.

IDENTIFICATION OF A FUNCTIONAL CTE-RELATED POST-TRANSCRIPTIONAL CONTROL ELEMENT IN INTRACISTERNAL A-PARTICLE RETROELEMENTS

Intracisternal A-particle retroelements (IAPs) are found in multiple copies in the genomes of rodents.[14] These retroelements have all characteristics of simple retroviruses such as LTRs, the putative genomic regions for *gag/pol* and *env*, the primer binding site and polypurine tract (Figure 3(a)). So far no intact IAP has been identified, since these elements are heavily mutated and, therefore, incapable of protein expression and propagation. Insertional mutagenesis occuring as a result of IAP integration is known to cause activation (by providing a promoter or polyadenylation signal) or disruption of the target gene.

Recently, we discovered a functional CTE-related element as part of a newly identified murine IAP, which is inserted within the transcribed mouse osteocalcin-related gene (ORG).[20] Analysis of this IAP genome revealed that the CTE-related element is located between the putative *env* gene and the 3' LTR, similar to the CTE location in the SRV genome (Figure 3(a)). This element shares several important features with the type D retrovirus CTEs such as the conservation of the two internal loops, preservation of their distance, and, to a lesser extent, the sequence of the hairpin loop (Figure 2). RNA sequence comparisons revealed that the similarity is confined mostly to the internal loop regions, whereas the predicted stem structures of this element show dramatic sequence divergence. The presence of a CTE-like element in the ORG-IAP is not unique to this retroelement. Several CTE-related elements have been subsequently found by us in other murine and rat IAPs. These elements show similarity only in the region of the internal loops of the CTE_{IAP}. Several of these elements contain nucleotide changes affecting their structure. These elements are also located between putative *env* coding region and the 3' LTR within the IAP retroelements. This finding demonstrates that the presence of CTE-related elements is a common feature of many retroelements.

Since retroelements such as the ORG-IAP have undergone severe mutations, including extensive deletions destroying the open reading frames, they are inactive. The question arises whether the identified CTE-related elements are still functional. Since no intact IAP is available, the function of the IAP CTE elements has been demonstrated using heterologous systems. A molecular clone of a type D retrovirus (SRV-1) is generated that lacks its CTE and cannot produce virus (Figure 3(b)). Upon insertion of the CTE from the ORG-IAP (CTE_{IAP}), it was shown that this element can replace the function of the CTE in SRV-1 and, hence, it is a functional posttranscriptional regulatory element.[20] This and other functional tests show that some of the other CTE-related elements found in other IAPs have shown impaired function or are inactive. Since these elements contain multiple nucleotide changes in the regions encompassing the consensus sequences, their impaired activity is expected.

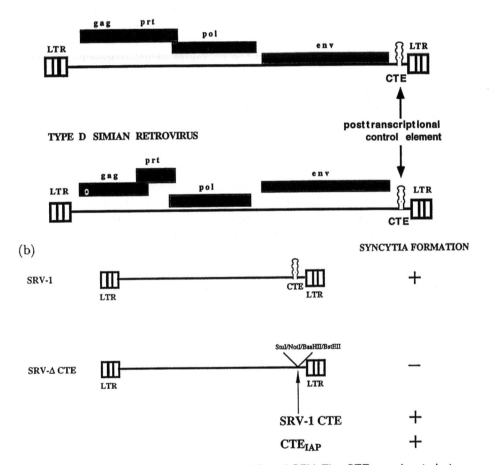

FIGURE 3 (a) Genomic organization of an IAP and SRV. The CTEs are located at similar positions between *env* and the 3' LTR. (b) The CTE is essential for SRV-1 expression. SRV ΔCTE has the CTE replaced by a polylinker. The indicated molecular clones were transfected into human 293 cells which were cocultivated with human Raji cells. Virus spread is monitored by syncytia assay. SRV ΔCTE does not produce any virus. Insertion of the SRV-1 CTE or the ORG-IAP results in virus production.

Taken together, the analysis of the SRV-1 CTE and the comparison of the SRV CTE and the CTE_{IAP}, strongly support a model in which the loop regions are the binding sites of cellular factors mediating CTE function (see above). Therefore, CTE_{IAP} and the SRV CTEs represent a novel class of RNA elements characterized by unique sequences within the internal loops, which are predicted to be interaction sites with cellular factors (Table 1).

The conservation of CTE-like elements in several IAP retroelements further suggests a *cis*-acting function essential for retrotransposition. By analogy to the type D retroviruses, it is thought that the CTE-like element promotes the export of the RNA from the nucleus to the cytoplasm where it is packaged. This is a necessary step allowing reverse transcription, followed by the integration of the retroelement in the host genome. Therefore, it is thought that the CTE together with the primer binding site, the polypurine tract and the packaging signal represents another essential *cis*-acting element necessary in the transposition process.

In addition to being important for the life cycle of the retroelement, insertional mutagenesis via IAP retroelement can introduce a CTE-like element into a cellular gene. As a consequence, the expression of the target gene could be affected at the posttranscriptional level. Analysis of the ORG-IAP revealed that it inserted between the promoter and the coding region of the osteocalcin-related gene, ORG.[20] ORG is a member of the murine osteocalcin gene cluster consisting of three genes, which are named osteocalcin 1, osteocalcin 2, and the osteocalcin-related gene. Although these genes have highly homologous promoters, it was observed that ORG is expressed in a developmental stage- and tissue-specific manner distinct from the expression of the two other osteocalcin genes (for detailed references see Tabernero et al.[20]). The osteocalcin genes 1 and 2 are expressed in the bone, whereas the osteocalcin-related gene is expressed in kidney. Analysis of the ORG transcripts in kidney revealed that it is still under the control of its promoter. Two species of transcripts are produced, one that contains the CTE and one that, as a result of terminal splicing, lacks CTE. Since these two transcripts have the same coding potential, the effect of CTE on their expression could not be determined. For the expression of the ORG, it was concluded that the location of the osteocalcin promoter in relation to the 5' LTR of the ORG-IAP and the acquisition of the CTE are most likely responsible for the distinct expression pattern of ORG.[20] Therefore, changes of the target gene expression at the posttranscriptional level is a novel concept in our thinking about effects of insertional mutagenesis via IAP.

In summary, the identification of functional CTE-related elements and their conservation in several intracisternal A-particle retroelements suggest a *cis*-acting function for the life cycle of the retroelement. Second, this offers the possibility that insertional mutagenesis via IAPs changes expression of rodent genes at the posttranscriptional levels. This regulatory feature adds an additional step in the complexity of altered gene expression as a result of insertion of an intracisternal A-particle retroelement.

AVIAN LEUKOSIS VIRUS EXPRESSION IS REGULATED POSTTRANSCRIPTIONALLY VIA SPECIES-SPECIFIC FACTORS

Avian leukosis-sarcoma viruses (ALV) also belong to the simple retroviruses encoding *gag/pol* and *env* (Figure 1), a genomic organization that is similar to that of the type D simian retrovirus. The unspliced mRNA encodes *gag/pol*, whereas the singly spliced mRNA encodes *env*. Some ALVs such as the Rous sarcoma virus (RSV) contain two direct-repeat (DR) elements located between *env* and the 3' LTR, but other ALVs contain only one element (Figure 1). The DR elements are present on both the unspliced and the singly-spliced transcripts. The DR elements can be folded into strong computer-predicted secondary structures (Figure 4). Recent findings have demonstrated that the DR elements are essential *cis*-acting elements, necessary for the production of the viral structural proteins. Therefore, like the type D retroviruses the ALV depend on posttranscriptional regulation for their expression. Whereas SRV utilizes ubiquitous primate cellular factors, the ALV production depends on avian-specific factors (Table 1). ALVs are replication-competent in avian cells, but produce only low levels of Gag and no infectious virions in mammalian cells (for detailed references see Nasioulas et al.[15]). One hypothesis has been that the production of the ALV structural proteins is controlled by a posttranscriptional regulatory mechanism that does not function in mammalian cells. Two lines of evidence support this model.

First, by introducing the HIV-1 Rev/RRE regulatory system (see below and Cullen, this volume), efficient expression and particle formation of ALV in mammalian cells was achieved.[15] The RRE was inserted in *cis* into an RSV-based vector. Upon transfection of mammalian cells, unspliced *gag/pol* mRNA was detected in the cytoplasm and Gag protein was produced efficiently, but only when an HIV-1 Rev expression plasmid was cotransfected. This finding demonstrates that the HIV-1 Rev/RRE regulatory system can counteract the factors responsible for the down-regulation of expression of ALV Gag protein in mammalian cells (Table 2). This observation supports the hypothesis that posttranscriptional regulation is responsible for the inability to produce ALV in mammalian cells, and that species-specific factors are responsible for expression in avian cells. Similar to the expression of the type D retrovirus, the ALV *env* gene is expressed independent of posttranscriptional regulation. These are two examples of *env* transcripts that contain a posttranscriptional control element but their expression is not regulated posttranscriptionally.

The second line of evidence that ALVs depend on posttranscriptional regulation was provided by the demonstration that the DR elements are essential for ALV expression in avian cells.[16] It was shown that at least one element is sufficient for replication of an RSV-based expression vector. It is unclear why some ALVs contain two elements. It was also shown that either of the DR elements can replace the Rev/RRE system to achieve HIV-1 Gag expression in chicken cells.

In summary, ALV expression is controlled at the posttranscriptional level. This regulation is mediated via the *cis*-acting DR elements that are thought to interact with putative avian-specific factors (Table 1). Therefore, both the type D retroviruses and the ALV depend on the cellular machinery for virus production. Whereas ubiquitous mammalian factors mediate SRV expression, avian-specific factors mediate ALV expression.

ALV DR

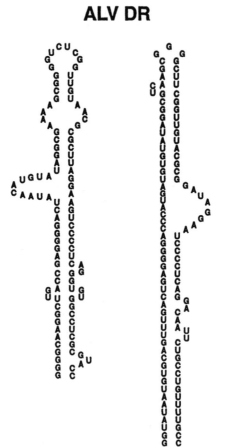

DR1   DR2

FIGURE 4 Computer-predicted structures of the DR elements of PrC RSV strain. Using the MFOLD program of the University of Wisconsin Genetics Computer Group package, the direct-repeat elements were folded. The strongest structures for DR1 and DR2 are shown, having a free energy of formation of −23.0 and −33.2 kcal/mol, respectively.

TABLE 2 Interchangeability of Viral Posttranscriptional Regulatory Systems.

Posttranscriptional Regulatory System	Retroviruses				Papillomaviruses
	SRV	ALV	HTLV	HIV/SIV	HPV/BPV
SRV CTE	+	nd[1]	+	+	+
IAP CTE	+	nd	nd	+	nd
Rex/RxRE	nd	nd	+	+	nd
Rev/RRE	+	+	+	+	+

[1] not done

POSTTRANSCRIPTIONAL CONTROL OF THE HTLV FAMILY OF ONCORETROVIRUSES

The HTLV family of oncoretroviruses consists of human T-cell leukemia virus type I (HTLV-I) and type II (HTLV-II), simian T-cell leukemia virus type I (STLV-I) and bovine leukemia virus (BLV). These viruses have the common retroviral genome organization consisting of the genes encoding Gag, Pol, and Env. In addition, these viruses have a region named X encoding the essential regulatory proteins Tax and Rex, as well as additional accessory proteins expressed by alternative splicing from short overlapping open reading frames (Figure 1) (for detailed references, see recent reviews by Franchini[10] and Gitlin[11]). Because these viruses produce additional regulatory and accessory proteins they are called complex retroviruses. Tax is a transcriptional activator of the LTR promoter, whereas Rex, analogous to Rev of HIV-1, regulates virus expression at the posttranscriptional level. Tax also participates in the transformation of the target cells. The accessory proteins are not required for HTLV-I and HTLV-II replication in cultured cells and their role in virus biology is still unclear. Infection of sheep with mutated BLV has shown that deletion of a nonessential part of the X-region results in low virus replication *in vivo* and no development of lymphoma in the infected sheep. It has been further shown that infection of rabbits with HTLV-II requires the complete X-region for efficient virus propagation. Although the accessory HTLV proteins are not essential for virus production *in vitro*, their conservation among the HTLV family of oncoretroviruses and the *in vivo* studies demonstrate that these accessory proteins may be important for virus-host interaction and may contribute to disease development. These findings are similar to those about the accessory proteins in HIV and SIV that are dispensable for virus propagation in most cell lines but are necessary for SIV pathogenicity in rhesus macaques (see Neuveut and Jeang, this volume).[7]

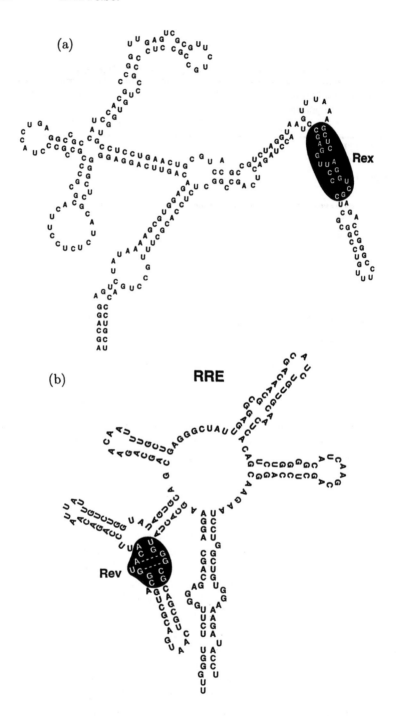

FIGURE 5   RNA structures of RXRE of HTLV-I (a) and RRE of HIV-1 (b) and the interaction site with Rex and Rev, respectively, are shown.

Rex regulates posttranscriptional events by binding to the RxRE. The RxRE has been shown to fold into a strong RNA secondary structure (Figure 5(a)) and is present in all viral mRNAs. Rex binds to a distinct region within the RxRE. It has been shown that Rex of HTLV-I and of BLV interact with the Rex responsive element (RxRE) located at the 3′ end of the transcript, whereas there is evidence that in HTLV-II, Rex binds to the RxRE located at the 5′ end of the transcript. However, in HTLV-II the 5′ RxRE is absent from the *env* mRNA due to the location of the major splice donor within this element, therefore, binding only to the 5′ RxRE element does not explain the mechanism responsible for the regulated expression of the *env* mRNA. The reason for the suggested difference between HTLV-I and BLV versus HTLV-II is not clear.

Rex is responsible for the nucleocytoplasmic transport of the both the unspliced *gag/pol* encoding mRNA and the singly spliced *env* encoding mRNAs. This is different from SRV and ALV, where *env* is expressed independently of posttranscriptional regulation. The other genes in the X-region such as *tax* and *rex* are expressed independently of Rex. Due to this posttranscriptional regulation pattern, expression of the viral mRNAs is divided into two phases, expression of the early genes encoding the regulatory and accessory proteins, and expression of the late genes encoding the structural proteins. It is possible that this complex regulation is necessary to ensure the production of all mRNAs.

For the transport of the Rex/RxRE-containing RNA complex to the cytoplasm, Rex is thought to interact with cellular proteins such as nucleoporins. A candidate nucleoporin-like protein RIP/RAB (see Cullen, this volume) has been identified. In addition, HTLV-I and HTLV-II produce the small accessory proteins p21x and x-III, respectively, from a transcript distinct from the doubly spliced mRNA encoding Rex. It was recently shown that these proteins affect the function of Rex by interfering with the intracellular trafficking and modification of Rex.[5,13] Therefore, these viruses seem to modulate their posttranscriptional regulation via positive- and negative-acting viral proteins.

In addition to representing the binding site for Rex, the RxRE has an additional function in the polyadenylation of all the viral transcripts. The RxRE overlaps part of the U3 and R region in the 3′ LTR. It has been noted that in these viruses polyadenylation occurs far away from the polyadenylation signal AAUAAA. The formation of the RxRE structure at the 3′ end of the transcripts brings the AAUAAA hexamer and the GU-rich elements in close proximity, resulting in proper 3′ processing of the poly(A) site.

The HTLV-I Rex protein does not only interact with the RxRE but it can also bind to the RRE element of HIV-1 (for a review see Felber[9]). The interaction site of Rex with the RRE is distinct from that of Rev. In contrast, Rev only interacts with the RRE and does not affect RxRE-containing mRNAs. Although Rex can substitute for Rev function resulting in the production of infectious virus, it does so with approximately 10% efficiency compared to Rev.

In summary, the HTLV oncoretroviruses have evolved to use a more complex regulatory system to control expression of the "late" mRNAs encoding *gag/pol* and *env.* These viruses utilize viral Rex protein that interacts with the *cis*-acting RxRE element to mediate posttranscriptional regulation.

LENTIVIRUS EXPRESSION DEPENDS ON THE VIRAL REV/RRE POSTTRANSCRIPTIONAL REGULATORY SYSTEM

The lentiviruses human immunodeficiency virus type 1 (HIV-1), human immunodeficiency virus type 2 (HIV-2), simian immunodeficiency virus type 1 (SIV), equine infectious anemia virus (EIAV), caprine arthritis encephalitis virus (CAEV), and visna virus are also complex retroviruses.[18] In addition to *gag/pol* and *env*, these viruses encode several other proteins: Tat and Rev are essential for the regulated expression of the virus at the transcriptional and posttranscriptional level, respectively, whereas the accessory proteins (Vif, Vpr/Vpx, Vpu, Nef) are essential in the virus life cycle *in vivo* but are dispensable for virus propagation in many cell lines *in vitro* (Figure 1). The overall arrangement of the genome organization is comparable in complexity to that of the HTLV oncoretroviruses. (See Introduction by Myers and Pavlakis, this volume.)

To express these different proteins, HIV-1, which is the best studied lentivirus, generates several mRNAs from the primary transcript through alternative splicing (for a review and additional references see Cullen (this volume),[6] Parslow,[17] and Pavlakis[18]). The various mRNAs can be divided into two groups: the unspliced and the intermediate-spliced mRNAs that contain the Rev-responsive element (RRE), and the multiply spliced mRNAs that lack the RRE. The unspliced and the intermediate-spliced mRNAs encoding Gag/Pol, Vif, Vpr, Vpu, and Env cannot be expressed in the absence of viral Rev protein. In contrast, there is no restriction for expression of the multiply spliced mRNAs encoding Tat, Rev, and Nef. Rev binds to a distinct region within the highly structured RRE-RNA element (Figure 5(b)) and is responsible for the nucleocytoplasmic export and expression of RRE-containing viral mRNAs. Therefore, HIV-1 replication depends on the presence of posttranscriptional activator protein Rev. Rev mediates its function by probably interacting with several cellular proteins exciting nuclear export function (see Cullen, this volume). The RRE is only present on the subset of viral mRNAs that depend on posttranscriptional regulation which is restricted to the mRNAs encoding the structural proteins. This is a clear difference from the arrangement found in the HTLV oncoretroviruses, the SRV and the ALV retroviruses, where the posttranscriptional control elements (RxRE, CTE, and DR, respectively) are present on all the viral mRNAs.

In the absence of Rev, the RRE-containing mRNAs are poorly expressed and no virions are produced. The identification of the molecular determinants which make a rescue by the Rev/RRE system necessary have been the target of intense studies. Several RNA elements in *gag/pol* and *env* that cause downregulation have been identified. The identified downregulatory elements are distinct from splice sites and act independently of splicing. Changes of the nucleotide composition without affecting the amino acid sequence led to successful elimination of several of these elements resulting mRNAs expressing *gag* or *gag/pol* independent of any posttranscriptional regulation. Recently, a cellular factor, poly-A binding protein 1, interacting with the downregulatory element located within the p17gag region has been identified[1] demonstrating that cellular mechanisms control the downregulation of the *gag/pol* and *env* mRNAs. The viral Rev/RRE system rescues these mRNAs resulting in their efficient expression. Therefore, Rev is also responsible for the balanced production of the different viral mRNAs, which is thought to be necessary for the production of infectious virions.

In summary, the lentiviruses also depend on posttranscriptional regulation which is mediated via the viral Rev protein that binds to the *cis*-acting RRE element. In contrast to the other retroviruses, only a subset of the viral mRNAs contains the *cis*-acting RRE. Rev has been shown to affect the fate of these mRNAs during nucleocytoplasmic transport and polysomal loading of the mRNAs. Rev, like Rex, utilizes the cellular export machinery for the transport of the viral mRNA from the nucleus to the cytoplasm. The detailed understanding of these mechanisms will also aid in the elucidation of posttranscriptional controls of cellular mRNAs.

VIRAL POSTTRANSCRIPTIONAL REGULATORY SYSTEMS ARE INTERCHANGEABLE

Table 1 summarizes the types of posttranscriptional regulation used by different retroviruses. In all cases, nucleocytoplasmic export of at least one type of viral RNA depends on this regulation. Although the detailed mechanisms of function of these processes may be different, the question arises whether these regulatory systems are interchangeable. As outlined below, the HIV-1 Rev/RRE system as well as the type D retrovirus CTE, which represents the binding site for a cellular "Rev-like" factor, are two potent regulatory systems that can be used to express heterologous posttranscriptionally regulated mRNAs (Table 2). These systems may be of general use in overcoming posttranscriptional restrictions due to divergent mechanisms, from nucleocytoplasmic transport to translation.

The Rev/RRE regulatory system has been shown to act on heterologous mRNAs (Table 2). Hybrid molecular clones of different viruses have been generated containing the HIV-1 RRE in *cis* and the Rev protein was provided in *trans*. It has been demonstrated that CTE-mediated regulation can be substituted by the HIV-1 Rev/RRE system.[8] Similarly, it has been shown that the species-specific restriction for ALV production in avian cells can be overcome in mammalian cells in the presence of Rev/RRE (see above).[15,16] Rev/RRE of HIV-1 can also replace the

Rex/RxRE system of the HTLV oncoretroviruses as well as the Rev regulation of several other lentiviruses (for a review see Felber[9]). The use of the Rev/RRE system has been important for the demonstration that expression of these retroviruses is controlled at the posttranscriptional level.

The CTE of the simian type D retroviruses has been shown to replace the Rev regulation (Table 2) of molecular clones of HIV-1 and SIVmac.[4,25,26] Bray et al.[4] have demonstrated that the presence of the MPMV CTE allows production of HIV-1 Gag/Pol and Env in the absence of Rev. Similarly, it has been shown that SIVmac Gag/Pol and a truncated Env protein could be produced from subgenomic vectors in the presence of the MPMV CTE.[19] We have demonstrated that the CTEs of SRV-1, SRV-2 and MPMV, when inserted into a Rev/RRE deficient molecular clone of HIV-1, can support continuous replication of the Rev-independent HIV-1 (Figure 6), resulting in the production of infectious virus in human peripheral blood mononuclear cells (PBMCs).[21,26] To prevent reversion to the wild-type genotype, these molecular clones contain multiple point mutations in *rev* and the RRE that do not affect the overlapping *tat* and *env* reading frames (Figure 6). Similarly, the Rev/RRE system of the infectious molecular clone SIVmac239 was replaced by the CTE resulting in propagation of Rev-independent SIV in rhesus PBMCs.[25] These Rev-independent HIV and SIV variants replicate to a lower extent in primary PBMCs and have lower infectivity than wild-type virus. The CTE identified in the mouse IAP is also able to replace Rev function, but it generates HIV variants that have more attenuated growth properties compared to the viruses containing the SRV CTEs.[20] These viruses can be propagated continuously for long periods of time without reversion to wild type or loss of the CTE.[21,25,26] Therefore, during the virus life cycle, Rev is required only as a component of the Rev/RRE positive regulatory system, and, once replaced by an appropriate alternative posttranscriptional mechanism such as the CTE, Rev is not essential for viral replication. Since expression of the Rev-independent viruses is mediated via cellular factors, the use of the CTE alters the temporal control of viral gene expression.

Another example of successful exchange of heterologous posttranscriptional control systems is the alleviation of the restricted expression of the capsid proteins of the papillomavirus by the HIV-1 Rev/RRE regulatory system and by the CTE in undifferentiated mammalian cells[2,22,23] (Table 2). This results in efficient L1 expression which is normally limited to the terminally differentiated keratinocytes (see also Baker, this volume).

In summary, the Rev/RRE system has been used to express the posttranscriptionally regulated mRNAs of SRV, ALV, HTLV and the papillomavirus L1 protein. CTE has been shown to support replication of Rev/RRE-deficient HIV-1 and SIVmac, to replace Rex function in an otherwise intact molecular clone of HTLV-I (J. Bear and Felber[3]) as well as to promote efficient expression of human papillomavirus L1 protein in undifferentiated mammalian cells. These two regulatory systems have many applications in replacing heterologous posttranscriptional control mechanisms.

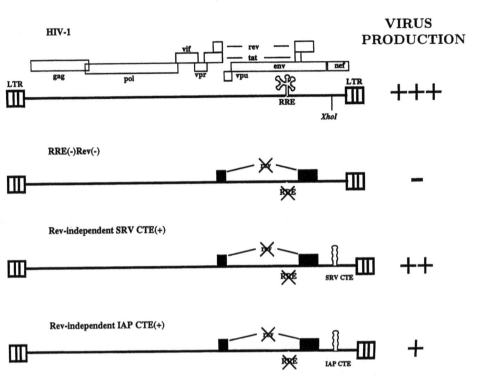

FIGURE 6   Rev/RRE deficient molecular clones of HIV-1 were generated that contain multiple point mutations in both *rev* and RRE without affecting the overlapping *tat* and *env* open reading frames. the SRV CTE and the IAP CTE were inserted into the *nef* open reading frame, respectively. Virus stocks were generated upon transfection of human 293 cells and used to infect human PBMCs. In the absence of *rev* and RRE no virus is produced.

RNA STRUCTURES MEDIATING VIRAL POSTTRANSCRIP-TIONAL REGULATION

The size and complexity of the different *cis*-acting RNA elements vary. The CTE and DR are simpler stem loop structures. Detailed analysis of the CTE has shown that the two internal loop regions represent the binding site for putative cellular factors and that most point mutations introduced into the element abolish its function. It is thought that the stem structures are crucial for proper positioning of the binding sites. Analysis of the CTE sequence from Rev-independent HIV-1 that propagated continuously for more than one year in PBMCs, showed that only rare nucleotide changes occured within the element, whereas the flanking regions

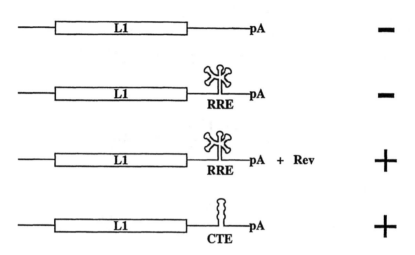

Expression in HeLa cells

FIGURE 7 Expression of the L1 mRNA is regulated at the posttranscriptional level. Upon transfection of undifferentiated HeLa cells with an L1 expression vector, no expression can be detected. Only in the presence of the HIV-1 RRE in *cis* and Rev in *trans*, or, alternatively, the presence of the SRV-1 CTE in *cis*, results in increased levels of cytoplasmic L1 mRNA and expression of high levels of L1 protein.

accumulated many insertions and deletions.[21] This confirms that nucleotide changes within the CTE are usually not tolerated and, if they occur, the virus having such an element will not be propagated. Little is known about the functional requirements of the DR elements. The RXRE is a more complex structure; this may be linked to additional functions of this element, representing both the interaction site of Rex as well as bringing polyadenylation site and polyadenylation signal into close proximity. The RRE folds also into a complex structure. Although the primary Rev binding site is an internal loop within stem loop II, deletions in the main stem far apart from the Rev interaction site abolish function. It was reported that the stem loop II structure containing the Rev binding site can mediate Rev binding and function, but multiple minimal elements were necessary for this to be accomplished. These findings suggest that the Rev binding site needs to be in a proper context and that additional Rev molecules and possibly other factors are required for function. Since RRE is contained within the *env* gene, mutations in RRE could also affect Env; this provides additional constraints on the tolerated mutations in this area. The RRE region lies within a conserved region in *env*.

ACKNOWLEDGMENTS

I thank the members of my lab J. Bear, C. Tabernero, A. von Gegerfelt, A. Zolo-tukhin as well as A. Valentin and G. N. Pavlakis for discussions and C. Rhoderick for assistance. Research sponsored by the National Cancer Institute, under contract with ABL.

REFERENCES

1. Afonina, E., M. Neumann, and G. N. Pavlakis. "Preferential Binding of Poly(A)-Binding Protein 1 to an Inhibitory RNA Element in the Human Immunodeficiency Virus Type 1 Gag mRNA." *J. Biol. Chem.* **272** (1997): 2307–2311.
2. Barksdale, S. K., and C. C. Baker. "The Human Immunodeficiency Virus Type I Rev Protein and the Rev-Responsive Element Counteract the Effect of an Inhibitory 5′ Splice Site in a 3′ Untranslated Region." *Mol. Cell Biol.* **15** (1995): 2962–2971.
3. Bear, J., and B. K. Felber. Unpublished results.
4. Bray, M., S. Prasad, J. W. Dubay, E. Hunter, K.-T. Jeang, D. Rekosh, and M.-L. Hammarskjold. "A Small Element From the Mason-Pfizer Monkey Virus Genome Makes Human Immunodeficiency Virus Type 1 Expression and Replication Rev-Independent." *Proc. Natl. Acad. Sci. USA* **91** (1994): 1256–1260.
5. Ciminale, V., L. Zotti, D. M. D'Agostino, and L. Chieco-Bianchi. "Inhibition of Human T-Cell Leukemia Virus Type 2 Rex Function by Truncated Forms of Rex Encoded in Alternatively Spliced mRNAs." *J. Virol.* **71** (1997): 2810–2818.
6. Cullen, B. R. "Mechanism of Action of Regulatory Proteins Encoded by Complex Retroviruses." *Microbiol. Rev.* **56** (1992): 375–394.
7. Desrosiers, R. C. "HIV with Multiple Gene Deletions as a Live Attenuated Vaccine for AIDS." *AIDS Res. Hum. Retroviruses* **8** (1992): 411–421.
8. Ernst, R. K., M. Bray, D. Rekosh, and M.-L. Hammarskjold. "A Structured Retroviral RNA Element that Mediates Nucleocytoplasmic Export of Intron-Containing RNA." *Mol. Cell Biol.* **17** (1997): 135–144.
9. Felber, B. K. "Regulation of mRNA Expression in HIV-1 and Other Retroviruses." In *Posttranscriptional Gene Regulation*, edited by D. Morris and J. B. Harford, 323–340. A Volume for the Series "Modern Cell Biology." New York, NY: Wiley, 1997.
10. Franchini, G. "Molecular Mechanisms of Human T-Cell Leukaemia/Lymphotropic Virus Type I Infection." *Blood* **86** (1995): 3619–3639.
11. Gitlin, S. D., J. Dittmer, R. L. Reid, and J. N. Brady. "The Molecular Biology of Human T-Cell Leukemia Viruses." In *Human Retroviruses*, edited by B. R. Cullen, 159–192. New York, NY: Oxford University Press, 1993.
12. Katz, R. A., and A. M. Skalka. "Control of Retroviral Splicing Through Maintenance of Suboptimal Splicing Signals." *Mol. Cell Biol.* **10** (1990): 696–704.
13. Kubota, S., M. Hatanaka, and R. J. Pomerantz. "Nucleo-Cytoplasmic Redistribution of the HTLV-I Rex Protein: Alterations by Coexpression of the HTLV-I p21x Protein." *Virology* **220** (1996): 502–507.
14. Kuff, E. L., and K. K. Lueders. "The Intracisternal A-Particle Gene Family: Structure and Functional Aspects." *Adv. Cancer Res.* **51** (1988): 183–276.

15. Nasioulas, G., S. H. Hughes, B. K. Felber, and J. M. Whitcomb. "Production of Avian Leukosis Virus Particles in Mammalian Cells Be Mediated by the Interaction of HIV-1 Rev and the Rev Responsive Element." *Proc. Natl. Acad. Sci. USA* **92** (1995): 11940–11944.

16. Ogert, R. A., L. H. Lee, and K. L. Beemon. "Avian Retroviral RNA Element Promotes Unspliced RNA Accumulation in the Cytoplasm." *J. Virol.* **70** (1996): 3834–3843.

17. Parslow, T. C. "Posttranscriptional Regulation of Human Retroviral Gene Expression." In *Human Retroviruses*, edited by B. R. Cullen, 101–136. Durham, NC; Oxford, New York; Tokyo: IRL Press at Oxford University Press, 1993.

18. Pavlakis, G. N. "The Molecular Biology of Human Immunodeficiency Virus Type 1." In *AIDS: Biology, Diagnosis, Treatment, and Prevention*, edited by V. T. DeVita Jr., S. Hellman, and S. A. Rosenberg, 45–74, 4th Ed. Philadelphia, PA: Lippincott-Raven, 1996.

19. Rizvi, T. A., R. D. Schmidt, K. A. Lew, and M. E. Keeling. "Rev/RRE-Independent Mason-Pfizer Monkey Virus Constitutive Transport Element-Dependent Propagation of SIVmac239 Vectors Using a Single Round of Replication Assay." *Virology* **222** (1996): 457–463.

20. Tabernero, C., A. S. Zolotukhin, J. Bear, R. Schneider, G. Karsenty, and B. K. Felber. "Identification of an RNA Sequence Within an Intracisternal-A Particle Element Able to Replace Rev-Mediated Posttranscriptional Regulation of Human Immunodeficiency Virus Type 1." *J. Virol.* **71** (1997): 95–101.

21. Tabernero, C., A. S. Zolotukhin, A. Valentin, G. N. Pavlakis, and B. K. Felber. "The Posttranscriptional Control Element of the Simian Retrovirus Type 1 Forms an Extensive RNA Secondary Structure Necessary for Its Function." *J. Virol.* **70** (1996): 5998–6011.

22. Tan, W., B. K. Felber, A. S. Zolotukhin, G. N. Pavlakis, and S. Schwartz. "Efficient Expression of the Human Papillomavirus Type 16 L1 Protein in Epithelial Cells by Using Rev and the Rev-Responsive Element of Human Immunodeficiency Virus or the *Cis*-Acting Transactivation Element of Simian Retrovirus Type 1." *J. Virol.* **69** (1995): 5607–5620.

23. Tan, W., and S. Schwartz. "The Rev Protein of Human Immunodeficiency Virus Type 1 Counteracts the Effect of an AU-Rich Negative Element in the Human Papillomavirus Type 1 Late 3′ Untranslated Region." *J. Virol.* **69** (1995): 2932–2945.

24. Valentin, A., C. Tabernero, and B. K. Felber. Unpublished observation.

25. von Gegerfelt, S. A., and B. K. Felber. "Replacement of Posttranscriptional Regulation in SIVmac239 Generated a Rev-Independent Infectious Virus Able to Propagate in Rhesus Peripheral Blood Mononuclear Cells." *Virology* (1997): in press.

26. Zolotukhin, A. S., A. Valentin, G. N. Pavlakis, and B. K. Felber. "Continuous Propagation of RRE(−) and Rev(−)RRE(−) Human Immunodeficiency Virus Type 1 Molecular Clones Containing a *Cis*-Acting Element of Simian

Retrovirus Type 1 in Human Peripheral Blood Lymphocytes." *J. Virol.* **68** (1994): 7944–7952.

Carl C. Baker
Basic Research Laboratory, Division of Basic Sciences, National Institutes of Health,
41 Library Dr. MSC 5055, Bethesda, MD 20892–5055; E-mail: ccb@helix.nih.gov

Posttranscriptional Regulation of Papillomavirus Gene Expression

INTRODUCTION

The papillomaviruses are small DNA tumor viruses which are epitheliotropic and productively infect only squamous epithelia. The life cycles are intimately linked with the differentiation program of the host cell, the keratinocyte. The viral genome is maintained as a low copy extrachromosomal plasmid in the undifferentiated or basal cells and only "early" viral genes are expressed in these cells. As the keratinocyte differentiates, vegetative viral DNA replication takes place in the spinous layer. Finally the "late" genes encoding the structural proteins are expressed in the fully differentiated cells of the granular layer and virus is produced. This differentiation-dependent life cycle serves two purposes, both of which allow the virus to set up a persistent infection which can shed virus over a relatively long period of time. Many DNA viruses take over host functions at late stages of the life cycle and eventually kill the cell in the process. In papillomavirus infection, late functions occur in cells which are destined to die anyway. In fact, cells in the late phase of the viral life cycle are constantly replenished from the population of persistently infected basal cells. Restriction of late gene expression to the superficial layers of a squamous epithelium may also help the virus escape detection by the immune system.

The genomes of the papillomaviruses are circular double-stranded DNAs approximately 8 kb in length. All papillomavirus genomes share a similar structural organization. All significant ORFs reside on one strand of the genome, and only this strand is expressed as mRNA (Figure 1(a)). Like other small DNA tumor viruses, the papillomavirus genomes can be divided into "early" and "late" regions based on the temporal order of expression. Unlike polyomavirus and SV40, whose early and late regions are expressed in a divergent manner (Figure 1(b)), the papillomavirus early and late mRNAs are expressed from one genome-length complex transcription unit. Multiple overlapping primary transcripts or pre-mRNAs are transcribed from multiple promoters in the early region as well as in a regulatory region upstream of the early region which has been named the long control region (LCR) or

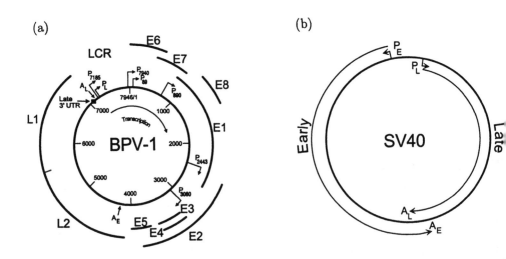

FIGURE 1 Maps of BPV-1 and SV40 genomes. (a) Circular map of BPV-1 genome. Open reading frames (ORFs) are indicated by arcs and designated early (E) or late (L). All transcription proceeds in the clockwise direction from promoters indicated by arrows and the label P_n, where n is the approximate start site for transcription. The late promoter is labeled P_L. The early and late poly(A) sites are indicated by arrows and the label A_E and A_L, respectively. LCR refers to the long control region which contains the origin of replication as well as transcriptional regulatory sequences. (b) SV40 map, showing divergent transcription of early and late regions from opposite strands of the DNA. The early and late promoters and early and late poly(A) sites are labeled P_E, P_L, A_E, and A_L, respectively.

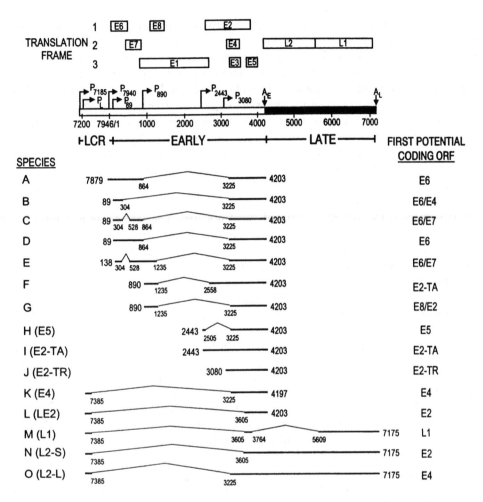

FIGURE 2 BPV-1 mRNAs. The top of the figure shows a linearized version of the
BPV-1 genomic map shown in Figure 1(a) and is annotated in a similar way except that
ORFs are indicated by open boxes. The locations of the early and late regions and the
LCR are shown below the map. At the bottom of the figure are the structures of some
of the mRNAs which have been identified in BPV-1 transformed C127 cells (species
A–J) and productively infected bovine fibropapillomas (species K–O). The numbers refer
to the first and last nucleotides of each exon. The identity of each mRNA is indicated at
the left and the first potential coding ORF at the right.

upstream regulatory region (URR). These pre-mRNAs are processed into a complex set of mRNAs through alternative splicing and alternative polyadenylation
(Figure 2). Splicing is the mechanism by which discontinuous regions of DNA can
be put together into one mRNA. Alternatively spliced mRNAs frequently can be

translated into different versions of a given protein. Alternative splicing may also control translation of downstream open reading frames in a polycistronic mRNA. In addition to being spliced, most polymerase II transcribed mRNAs have defined 3′ ends which are generated by polyadenylation, a process involving endonucleolytic cleavage at a poly(A) site followed by addition of a poly(A) tail. The pre-mRNA downstream of the poly(A) site is generally degraded after polyadenylation occurs. Transport of an mRNA from the nucleus to the cytoplasm appears to be linked with polyadenylation in most cases. In addition, the poly(A) tail plays important roles in the regulation of mRNA stability and translation. Separate poly(A) sites lie at the 3′ ends of the papillomavirus early and late regions and have been named the early and late poly(A) sites, respectively (Figure 2). The early poly(A) site is probably used at all stages of the viral life cycle, whereas use of the late poly(A) site is restricted to fully differentiated keratinocytes. The late genes encoding the major (L1) and minor (L2) capsid proteins lie between these two poly(A) sites. Therefore, an mRNA encoding these two genes is made only when polyadenylation takes place at the second (late) poly(A) site.

Transcription of papillomavirus genomes as a complex transcription unit allows the expression of multiple, overlapping genes to be differentially regulated through both transcriptional and posttranscriptional mechanisms. In this chapter I will discuss the regulation of complex transcription units as it applies to the papillomaviruses. There will be an emphasis on posttranscriptional regulation since transcriptional regulation is discussed elsewhere in this book (see McBride et al. and Thierry and Yaniv, this volume). The bovine papillomavirus type 1 (BPV-1) will be the major model system, although other papillomaviruses will also be discussed. Although the details may differ between the papillomaviruses, the general principles of posttranscriptional regulation in a complex transcription unit will hold for all the papillomaviruses. In addition, parallels will be made with HIV-1 and with several other viruses and cellular genes.

Several general principles can be deduced from the studies described below and apply equally to the papillomaviruses as well as to many other viral and cellular genes. Regulation of a gene is frequently very complex and involves both transcriptional and posttranscriptional mechanisms. This allows the levels of expression to be responsive to several different input signals. In the case of the papillomaviruses, regulation probably takes place at every conceivable level. While any single mechanism may modulate expression levels by only a relatively small amount, the overall effect can be quite large. Alternative splicing and polyadenylation of a pre-mRNA involve multiple, mutually exclusive processing events that compete with each other. This requires a fine balance between processing pathways that is accomplished through finely tuned suboptimal processing signals whose activity is further modulated by multiple positive and negative *cis* elements. In addition, factors from different processing pathways can interact with each other and these interactions can have both positive and negative effects on a given processing event.

TRANSCRIPTIONAL PAUSING AND TERMINATION

Processing of a pre-mRNA is thought to be a cotranscriptional event. Therefore, the order in which processing sites are transcribed and the rate of transcription can significantly affect the pattern of processing when there are multiple competing processing choices that are mutually exclusive. The simplest example of this is two tandem poly(A) sites (Figure 3). In the case of two sites that are close together and therefore transcribed at approximately the same time, the stronger or more efficient site (see below) will be used preferentially (Figure 3(a)). A delay in the transcription of the distal site will favor processing at the proximal site, since only this site is initially available for processing (Figure 3(b)). One way to delay transcription of the second site is to increase the distance between the two sites. The rate of transcription of eukaryotic genes has been estimated to be 1 kb/minute. At this rate it would take approximately 3 minutes to transcribe from the early to the late poly(A) site of BPV-1. This factor alone would favor use of the BPV-1 early poly(A) site. A second way to delay transcription is through transcriptional pausing. Transcriptional pausing is a poorly understood phenomenon that involves a roadblock to transcriptional elongation. This can be due to structural features of the DNA (such as bent DNA) or to the presence of a DNA binding protein. Ashfield et al.[2] demonstrate that a pause site placed between a weak poly(A) site and a downstream strong poly(A) site increases usage of the upstream weak poly(A) site. An extreme example of this is transcription termination, which would prevent the second poly(A) site from ever being transcribed. Transcription termination between the early and late poly(A) sites has been demonstrated using nuclear run-off assays in BPV-1 transformed mouse cells.[3] In addition, *in situ* hybridization studies of tissue productively infected by HPV-6 show more nuclear RNA hybridizing to L2 ORF probes than to L1 ORF probes, suggesting that there is also transcription termination between the early and late poly(A) sites of HPV-6.[24] The prevailing theory of transcription termination downstream of most *pol* II-transcribed genes is that termination is linked to cleavage at an upstream poly(A) site (reviewed in Proudfoot[22]). Basically, cleavage at that site exposes the uncapped 5' end of the downstream nascent transcript to degradation by an exonuclease. Termination is thought to occur when the exonuclease reaches the RNA polymerase while it is stalled at a pause site. If this theory is correct, termination within papillomavirus late regions would take place only after a choice of poly(A) sites has been made and therefore would have no regulatory significance.

The role of transcriptional pausing in the regulation of papillomavirus poly(A) site selection has not been studied. However, pausing could also affect processing when a poly(A) site lies in an intron (Figure 3(b)). This is the situation in papillomavirus pre-mRNAs where the early poly(A) site is in the second intron of the L1 mRNA (Figure 2). Even a strong poly(A) site is not used when located

(a) **Conditions that favor use of the promoter-distal poly(A) site (A₂):**

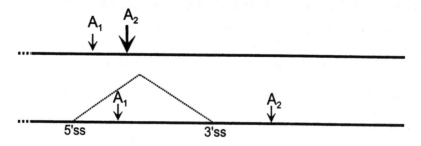

(b) **Conditions that favor use of the promoter-proximal poly(A) site (A₁):**

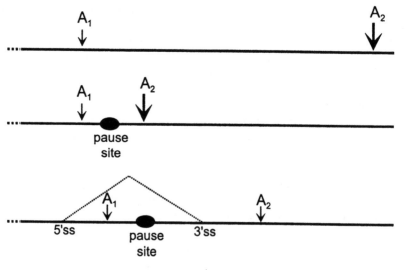

FIGURE 3   Factors which affect choice of alternative poly(A) sites. (a) Conditions which favor use of the promoter-distal poly(A) site. In the first example, the larger arrow indicates that A₂ is the stronger poly(A) site and is therefore favored. In the second example, A₁ is not used because it is in an intron. The 5' and 3' splice sites are labeled 5'ss and 3'ss, respectively. (b) Conditions which favor use of the promoter-proximal poly(A) site. Increasing the distance or pausing between two poly(A) sites delays transcription of the downstream poly(A) site and favors use of the upstream poly(A) site, even if it is the weaker site. Transcriptional pausing between a poly(A) site and downstream 3' splice site can activate a poly(A) site located in an intron.

in an intron because splicing predominates over polyadenylation (Figure 3(a)) (see Enriquez-Harris et al.[15] and references therein). A trivial explanation for this observation is that once an intron is removed, the poly(A) site is no longer available in the pre-mRNA and therefore can't be used. However, competition between splicing and polyadenylation may also involve the competition between assembly of mutually exclusive commitment complexes. A commitment complex is the first stable complex in the assembly of the spliceosome or poly(A)osome and "commits" the pre-mRNA to being processed by the corresponding pathway. Thus, transcriptional pausing between the poly(A) site and the 3′ splice site will favor polyadenylation over splicing by delaying the assembly of a spliceosome on the pre-mRNA.[15] Thus, transcriptional pausing within the 5′ end of the papillomavirus late region at early stages of the viral life cycle could be one mechanism which ensures that only the early poly(A) site is used at this time.

DIRECT REGULATION OF POLYADENYLATION

The strength or efficiency of a poly(A) site determines how well a site will compete with other poly(A) sites or with splicing. Strength appears to be a function of the overall stability of the polyadenylation complex, which is determined, at least partially, by the affinity of the cellular polyadenylation factor CstF for a downstream GU/U-rich element (reviewed in Wahle and Keller[28]). The BPV-1 early poly(A) site does not have a good looking downstream element (DSE), suggesting that this site is inherently weak. In contrast, the BPV-1 late poly(A) site has a good DSE, suggesting that this site is inherently strong and should be used preferentially over the early poly(A) site if both are present in the pre-mRNA at the same time. Surprisingly, the suboptimal early poly(A) site is the only poly(A) site used at early stages of the viral life cycle even though it is located in an intron (Figure 2). As mentioned above, even a strong poly(A) site isn't used when located in most introns. One reason polyadenylation at the suboptimal early poly(A) site is able to dominate over splicing is that the 5′ splice site at nt 3764 is also suboptimal, having a GC dinucleotide instead of the canonical GU dinucleotide common to most 5′ splice sites. This arrangement balances weak splicing with weak polyadenylation and, thus, allows regulation of the two processes. Since it is only necessary to regulate one of the two processes, the question is which process is regulated.

A clue comes from the following experiments done in BPV-1 transformed mouse cells where processing of BPV-1 pre-mRNAs normally uses only the early poly(A) site. Mutation of the BPV-1 early poly(A) site does not lead to significant use of the late poly(A) site.[1] Instead, a weaker cryptic poly(A) site located approximately 100 nt upstream of the early poly(A) site is used predominantly. If this cryptic early poly(A) site is also mutated, polyadenylation takes place at additional cryptic sites in the same general region. Similar results have also been obtained with a deletion that included the early poly(A) site.[10] These results have also been reproduced in transient expression assays using a vector expressing a BPV-1 late pre-mRNA.[4] Thus, it appears that there are very powerful mechanisms which prevent the late poly(A) site from functioning at early stages of the viral life cycle. A very different result is obtained in the transient assay system if the weak 5' splice site at nt 3764 is mutated to a strong consensus site. In this situation, all early poly(A) sites are bypassed and splicing predominates over polyadenylation.[4] While it is difficult to conceive of a specific polyadenylation regulatory element which would down-regulate polyadenylation at the early poly(A) site as well as all cryptic early sites at late stages of the viral life cycle, splicing is able to do this. This suggests that splicing may be the primary regulated process.

Another possibility is that papillomavirus poly(A) site choice is regulated by levels of general cellular polyadenylation factors. Under conditions in which the levels or activity of one or more general polyadenylation factors is low, the weak early poly(A) site (and the cryptic early sites) would compete poorly for these factors, and polyadenylation at these sites would be slow. This would give more time for RNA polymerase to transcribe the stronger late poly(A) site which would then compete more effectively than the early poly(A) site for polyadenylation factors. It is not known if differentiated keratinocytes have low levels of polyadenylation factors. However, there are examples of complex transcription units where poly(A) site choice is regulated in this way. The IgM heavy chain gene has two poly(A) sites: the proximal (μs) site is used to make the mRNA for the secreted form, while the distal site (μm) is used for the membrane-bound form (Figure 4). The μs poly(A) site lies in an intron flanked by a suboptimal 5' splice site, so splicing and polyadenylation are balanced (reviewed in Proudfoot[21]). The polyadenylation factor CstF binds more strongly to the μm site than to the μs site. Furthermore, levels of the 64 kDa subunit of CstF (CstF-64) are low in mouse primary B cells, which express predominantly the membrane-bound IgM. Overexpression of CstF-64 is sufficient to switch heavy chain expression from the membrane-bound form to the secreted form, suggesting that CstF-64 plays an important role in regulating IgM heavy chain expression during B cell differentiation. A similar situation exists during adenovirus infection (reviewed in Proudfoot[21]). The adenovirus major late transcription unit contains five poly(A) sites. At early times in infection, the first (L1) poly(A) site is used three times more frequently than the third (L3) poly(A) site. The opposite is true late in infection when L3 is used three times more frequently than L1. As in the IgM example, the distal (L3) poly(A) site binds CstF more strongly than the proximal (L1) site. At late times during infection, the activity (but not the concentration) of CstF decreases. Under these conditions the L3 poly(A) site competes more effectively for CstF and therefore is used preferentially.

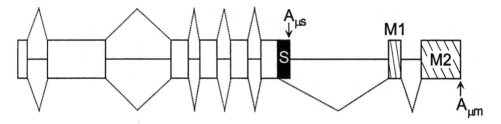

FIGURE 4   Alternative processing of the IgM heavy chain pre-mRNA. The processing pattern for the secreted form is shown above the pre-mRNA while the pattern for the membrane-bound form is shown below. The gray boxes are the constitutively spliced exons. The alternative exons for the secreted form and the membrane-bound form mRNAs are solid and hatched boxes labeled S, M1, and M2, respectively. The poly(A) sites for the secreted form and the membrane-bound form mRNAs are labeled $A_{\mu s}$ and $A_{\mu m}$, respectively.

REGULATION OF BPV-1 ALTERNATIVE SPLICING

Most BPV-1 pre-mRNAs are alternatively spliced, yielding a large number of different transcripts with different coding capacities (Figure 2). The majority of spliced early mRNAs as well as several late mRNAs are made using the 3′ splice site at nt 3225. At late stages of the viral life cycle, a novel 3′ splice site at nt 3605 is used to generate the L1 (major capsid) mRNA as well as an L2 mRNA (L2-S).[5] *In situ* hybridization studies of bovine warts have demonstrated that activation of this 3′ splice site is restricted to the differentiated granular layer, indicating that alternative splicing is differentiation dependent.[7] The importance of this switch in 3′ splice site utilization is that it may be the key regulatory mechanism which controls the early to late shift in processing of the late pre-mRNA. Basically, size restrictions on internal exon recognition may link activation of the suboptimal nt 3764 5′ splice site (and, therefore, subsequent removal of intron 2 of the L1 mRNA) with activation of the nt 3605 3′ splice site[5] (see Figure 2 and 5). This can be explained by the exon definition theory of splicing, which suggests that recognition of an internal exon (such as L1 exon 2) in vertebrates is restricted to 300 nt or less (reviewed in Berget[8]). These restrictions on internal exon size are due to cooperative interactions which span the exon and involve splicing factors bound at the 5′ and 3′ splice sites. Restrictions on internal exon size are even more pronounced when one of the splice sites flanking the exon is suboptimal, as is the case with the nonconsensus 5′ splice site at nt 3764 (Figure 5(c)). Thus, the L1 exon 2 from the 3′ splice site at nt 3605 to the 5′ splice site at nt 3764 is only 160 nt long and, therefore, can be efficiently recognized, while a hypothetical alternative internal exon from nt 3225 to 3764 would be 540 nt long and, therefore, inefficiently recognized. Furthermore, alternative splicing may be indirectly linked to alternative polyadenylation since activation of the suboptimal nt 3764 5′ splice site may be required to bypass the early poly(A) site and allow the late poly(A) site to be used (see above).

We are beginning to understand the regulation of alternative splicing of the BPV-1 late pre-mRNA. Important determinants of alternative splicing include not only the splice sites themselves but also multiple *cis*-acting elements distinct from the splice sites. In this way the regulation of papillomavirus alternative splicing is similar to that of HIV-1. This will be described more fully below. Analysis of the BPV-1 nt 3225 and nt 3605 3′ splice sites suggests that both are suboptimal (Figure 5(b)). This is to be expected since strong processing sites are difficult to regulate. Both sites have purines interrupting the polypyrimidine tracts and would

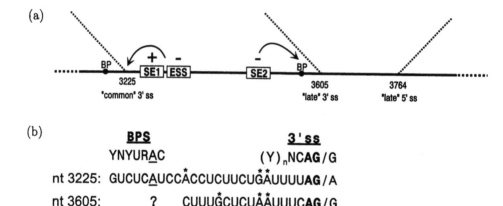

FIGURE 5   BPV-1 splice sites and splicing regulatory elements. (a) Model for regulation of BPV-1 alternative 3′ splice site selection. SE1 and the ESS enhance or suppress use of the nt 3225 3′ splice site, respectively. SE2 represses use of the nt 3605 3′ splice site. Branch point (BP). (b) The sequences of the nt 3225 branch point (BPS) and the sequences of the nt 3225 and 3605 3′ splice sites (3′ss) are shown along with the BPS and 3′ splice site consensus sequences. The nt 3605 BPS has not been identified. The BP adenosine is underlined. Purines in the polypyrimidine tract are indicated by an *. Y, pyrimidine; R, purine; N, any nucleotide. (c) The sequence of the nonconsensus nt 3764 5′ splice site (5′ss) is shown along with the 5′ splice site consensus sequence. M refers to A or C.

not be expected to have a strong affinity for the splicing factor U2AF[65]. In addition, the nt 3225 3' splice site has a suboptimal branch point sequence (BPS) and is used inefficiently *in vitro*.[29] The BPS is part of a 3' splice site and is defined in part by base pairing interactions with the U2 small nuclear RNA (U2 snRNA) component of the U2 small nuclear ribonucleoprotein particle (U2 snRNP). An A residue in the BPS makes a nucleophilic attack on the 5' splice site in the first step of splicing. Several *cis*-acting splicing elements have now been identified which regulate these alternative splice sites.[29] Immediately downstream of the nt 3225 3' splice site is a bipartite element which can both positively and negatively regulate this site (Figure 5(a)). This element consists of a purine-rich exonic splicing enhancer (SE1) followed by a pyrimidine-rich exonic splicing suppressor (ESS) (Figure 5(a) and 6). Exonic enhancers have been shown to stimulate splicing by recruiting U2AF to suboptimal polypyrimidine tracts at the 3' splice site (reviewed in Manley and Tacke[20]). In the case of purine-rich exonic splicing enhancers, this function is mediated by a family of splicing factors known as SR proteins. The ESS may function by competing with the polypyrimidine tract at the nt 3225 3' splice site for the binding of U2AF and perhaps other splicing factors. A second purine-rich splicing enhancer-like element (SE2) has also been found a short distance upstream of the nt 3605 3' splice site. Both SE1 and SE2 bind several SR proteins including ASF/SF2.[30] Although SE2 can function as a splicing enhancer when placed in an exonic location in a heterologous pre-mRNA, its role in the splicing of BPV-1 the late pre-mRNA is not known. SE2 may be too distant (approximately 250 nucleotides) from the nt 3225 3' splice site to act as an enhancer for this site (discussed in Manley and Tacke[20]). Recent data on the regulation of adenovirus alternative splicing suggests that SE2 may function as an intronic splicing suppressor and inhibit use of the nt 3605 3' splice site (reviewed in Manley and Tacke[20]). SR protein binding to a purine-rich intronic element inhibits Adenovirus IIIa splicing by blocking the binding of the U2 snRNP at the branch point. Like BPV-1 SE2, the adenovirus purine-rich element can also function as an ESE when located in an exon.

The following model is suggested for the regulation of BPV-1 alternative 3' splice site selection (Figure 5(a)). The arrangement of ESE and ESS elements between two 3' splice sites allows the coordinated regulation of two alternative splice sites by the same transacting factors. High SR protein binding activity would enhance use of the nt 3225 3' splice site through binding to SE1, and repress use of the nt 3605 3' splice site through binding to SE2. If SR protein binding activity decreases, then the activity of the ESS would dominate, repressing use of the nt 3225 3' splice site. Coupled with this would be a loss of nt 3605 suppression due to SE2. The net result would be a shift in 3' splice site utilization. This, perhaps coupled with a facilitated recognition of the suboptimal nt 3764 5' splice site would suppress use of the early poly(A) site, allowing the late poly(A) site to be utilized and the late genes to be expressed. One interesting feature from the virology standpoint is that so far there is no evidence that the virus plays an active role in regulating alternative splicing. It is likely that the papillomaviruses have evolved to take advantage of cellular regulatory pathways involved in the alternative splicing of cellular genes during keratinocyte differentiation.

Splicing regulatory elements with similar functions have been identified in a number of viral and cellular pre-mRNAs. Although exonic splicing enhancers appear to be relatively common, only a few exonic splicing suppressors have been identified. In HIV-1, balanced splicing is required to produce not only the multiply spliced mRNAs encoding proteins such as *tat, rev,* and *nef,* but also the unspliced and singly spliced genomic RNA and mRNAs encoding *env, gag,* and *pol.* This balanced splicing requires both suboptimal 3′ splice sites as well as additional positive and negative exonic elements which regulate these suboptimal sites. The *tat/rev* exon 3 contains a bipartite splicing regulatory element which is similar to the BPV-1 SE1/ESS element in that it consists of both an exonic splicing enhancer (ESE) and an exonic splicing silencer (ESS) (see Si et al.[23] and references therein). An ESS has also been identified in the HIV-1 *tat* exon 2. Inclusion or exclusion of the human fibronectin EDA exon is also regulated by a bipartite splicing element.[11] A final example of an

BPV-1 SE1   GAAGGACCUGAAGGAGACCCUGCAGGAAAAGAAGCCGAGCCAGCCCAGCC

BPV-1 SE2   AGGCAGGAAGAAGAGGAGCAGUCGCCCGACUCCACAGAGGAAGAACCAG

BPV-1 ESS   UGUCUCUUCUUUGCUCGGCUCCCCCGCCUGCGGUCCCAUCAGAGCAGGC

HIV-1 *tat* exon 2 ESS   AGAUCCUAGACUAGAGCCCU

HIV-1 *tat* exon 3 ESS   AGAUCCAUUCGAUUAGUGAA

HIV-1 *tat* exon 3 ESE   GAAGAAGAAGGUGGAGAGAGAGACAGAG

h fibronectin EDA exon ESE   GAAGAAGAC

h fibronectin EDA exon ESS   CAAGG

FGFR-2 K-SAM exon ESS   UAGG

FIGURE 6   Sequences of exonic splicing enhancer (ESE) and exonic splicing suppressor (ESS) elements. Solid underlines indicate sequence similarities between the different ESEs. Dashed underline indicates sequence similarity between the two HIV-1 ESS elements. The gray box indicates the 10 nt core HIV-1 ESS element, which is necessary and sufficient for splicing suppression.

exonic sequence that inhibits splicing has been identified within the alternative K-SAM exon of the fibroblast growth factor receptor-2 gene.[13] The sequences of these splicing regulatory elements are summarized in Figure 6. All splicing enhancers share sequences similar to the fibronectin enhancer (GAAGAAGA). This sequence is similar to the consensus binding site (RGAAGAAC) for the SR protein ASF/SF2 determined by SELEX.[25] SELEX is an *in vitro* selection procedure in which a randomized RNA sequence is bound to an RNA binding protein and the "winners" are amplified by RT-PCR and reexpressed as RNA. Optimal binding sites can be derived by sequencing the final product after multiple rounds of selection. SR proteins including ASF/SF2 have been shown to bind to several purine-rich ESEs including the BPV-1 SE1 and SE2 (reviewed in Manley and Tacke[20]; Z. M. Zheng, P. He, and C. C. Baker[30]), and these proteins are thought to mediate the function of these elements. In contrast, there are no obvious similarities between the sequences of the splicing suppressors from the different systems. This suggests that they bind different factors and function by different mechanisms.

Although most BPV-1 mRNAs are spliced, there are a few notable exceptions (Figure 2). Two examples are the E2-TA mRNA (transcribed from P_{2443} and encoding the full-length E2 transcriptional activator protein) and the E2-TR mRNA (transcribed from P_{3080} and encoding the C-terminal E2 transcriptional repressor). Unspliced mRNAs in other systems such as HIV-1, type D retroviruses, and HSV require special *cis* elements and transacting viral or cellular proteins for their expression (see Felber and also Cullen, this volume). It is not yet known if this is also the case for the unspliced BPV-1 mRNAs. The pre-mRNA transcribed from P_{2443} is predominantly spliced to provide the mRNA from which the E5 oncoprotein is translated (Figure 2). Thus, regulation of the splicing of this pre-mRNA could control E5 and E2-TA protein levels. Since this pre-mRNA is spliced using the nt 3225 3' splice site, the same *cis* elements that regulate 3' splice site selection in the BPV-1 late pre-mRNA could be involved.

3'UTR ELEMENTS

A variety of posttranscriptional regulatory elements are found in the 3' untranslated regions (3'UTRs) of viral and cellular genes and influence expression through effects on polyadenylation, nucleocytoplasmic transport, mRNA stability, and translation. This is also true for the papillomaviruses. In several cases, the elements are complex and may bind multiple factors and function through multiple mechanisms. The late 3'UTR of BPV-1 contains a regulatory 5' splice site, which inhibits expression of the L1 and L2 mRNAs but is not used for splicing[16] (Figure 7(a)). The identity of this element as a 5' splice site was confirmed by a genetic suppression assay, which demonstrated that the function of the element is dependent on the binding of the U1 snRNP, a splicing factor which normally recognizes 5' splice sites. The mechanism by which this 5′ splice site inhibits expression is unknown, but could be through either inhibition of polyadenylation or by functioning as a nuclear retention signal. A 5′ splice site upstream of a poly(A) site has been shown to increase levels of nuclear pre-mRNA that has not been cleaved and polyadenylated.[4] This observation is most consistent with inhibition of polyadenylation. *In vitro* studies have also shown that a 5′ splice site inhibits polyadenylation.[12] In polyomavirus, a 5′ splice site has been shown to retain mRNAs in the nucleus, indicating that a 5′ splice site can act as a nuclear retention signal.[17] Also consistent with this mechanism is the observation that the HIV-1 Rev protein in *trans* and the Rev binding site, the RRE, in *cis* can block the effect of an inhibitory 5′ splice site.[6] The Rev protein is thought to facilitate the nucleocytoplasmic transport of HIV-1 unspliced and singly spliced transcripts and contains a nuclear export signal which is essential for this effect (see Cullen and also Felber, this volume). One possible model, which is compatible with all the data, is that when Rev is bound to the RNA, the RNA enters a transport pathway where the pre-mRNA has access to the polyadenylation machinery but not the splicing machinery.

(a) BPV-1 late 3′UTR

U1   3′...GACGGUCCAUUCAU$_m$A$_m$pppG^{m_3} 5′

mRNA   5′___**AAGGUAAGU**___11 nt___AAUAAA___ 3′

(b) HPV-16 late 3′UTR

5′__GCUAAACGCAAAAAACGUAAGCUGUAAGUAUUGUAUGUAUGUU

GAAUUAGUGUUGUUUGUUGUGUAUAUGUUUGUAUGU__112 nt__AAUAAA__ 3′

(c) HPV-1 late 3′UTR

5′___**AUUUA**__10 nt__**AUUUA**__7 nt__(UUUUUAUA)$_3$__368 nt__AAUAAA___ 3′

FIGURE 7 Negative posttranscriptional regulatory elements in papillomavirus late 3′ untranslated regions (3′UTRs). (a) The BPV-1 element (bold) is a regulatory 5′ splice site which base pairs with the 5′ end of the U1 snRNA. (b) The HPV-16 element has 4 overlapping nonconsensus 5′ splice site-like sequences (underlined) followed by a GU-rich region (Gs and Us shown in bold). (c) Two AUUUA sequences (bold) are essential for activity, of the HPV-1 late 3′UTR, although three UUUUUAUA sequences also contribute to full activity of the element. AAUAAA is the polyadenylation signal.

The late 3'UTR of HPV-16 also contains a negative posttranscriptional regulatory element which has some similarities with the BPV-1 late 3'UTR element (Figure 7(b)) (see Dietrich-Goetz et al.,[14] Furth et al.,[16] and references therein). Part of this element has four overlapping weak (nonconsensus) 5' splice sites of which no single site is sufficient for the inhibitory activity. Since two consensus 5' splice sites act synergistically to inhibit expression, it is likely that splicing factors bind cooperatively to the adjacent sites.[6] Downstream of the four weak 5' splice sites is a GU-rich inhibitory element, which binds a 65 kDa nuclear protein that may be the splicing factor U2AF^{65}. Significantly, inhibition by the combined region is relieved by treatment of keratinocytes with PMA, an agent which induces differentiation. This suggests that this element may play an important role in regulating the differentiation-dependent expression of HPV-16 late genes. The mechanism(s) by which this element functions is not known. Although the element has been shown to destabilize RNA *in vitro*, this effect has not been demonstrated *in vivo*. The element has also been shown not to affect polyadenylation *in vitro*. However, not all polyadenylation elements reproducibly affect polyadenylation *in vitro*. Like the BPV-1 element, interaction of the 5' splice sites and GU-rich regions with splicing factors could inhibit either polyadenylation or nucleocytoplasmic transport. An alternative possibility is suggested by the fact that GU-rich and U-rich elements downstream of the poly(A) site bind a polyadenylation factor CstF which enhances polyadenylation (reviewed in Wahle and Keller[28]). It is possible that the GU-rich region inhibits polyadenylation by competing for CstF binding or that the binding of CstF upstream of the poly(A) site might form a complex which is incompetent for polyadenylation. Finally, it should be noted that the *in vivo* expression studies cited above only looked at the levels of the reporter protein chloramphenicol acetyl transferase (CAT) and, therefore, could not distinguish between an effect on translation versus an effect on mRNA levels. Tan et al.[26] have carried out *in vivo* expression studies using an HIV-1 p17^{gag} reporter and have shown that the HPV-16 3'UTR element inhibited translation much more strongly than it decreased steady state levels of mRNA. It appears that this element may function at multiple levels. Interestingly, the HIV-1 Rev protein is able to reverse the effects of this element on both mRNA and protein levels as long as Rev is tethered to the RNA through the RRE. This is a good demonstration of Rev's ability to influence multiple posttranscriptional regulatory pathways.

The presence of inhibitory elements in papillomavirus late 3'UTRs may be a general feature of these viruses. Yet another inhibitory element is present in the HPV-1 late 3'UTR[27] (Figure 7(c)). This element also consists of two parts: the most important part of the element contains two AUUUA motifs while three UU-UUUAUA motifs are less important. The function of this element was investigated using an HIV-1 p17^{gag} reporter. The major effect of this element appears to be on the efficiency of translation, while having only a modest effect on cytoplasmic mRNA levels. The mechanism of action has not been determined. However, AU-UUA sequences in the 3'UTRs of cellular genes have been shown to destabilize mRNAs and inhibit translation. Like the BPV-1 and HPV-16 late 3'UTR elements, the activity of the HPV-1 late 3′UTR element can also be relieved by the HIV-1 Rev protein. Tan and Schwartz[27] suggest that mRNAs containing this element enter an unproductive pathway in the nucleus leading to degradation and inefficient translation. The Rev protein presumably directs the RNAs to a productive nuclear export pathway.

Papillomavirus early 3′UTRs may also contain posttranscriptional negative regulatory elements. The HPV-16 E6/E7 mRNA is unstable with a half-life of only 20–40 minutes in NIH 3T3 cells.[18] This instability is most likely due to elements in the early 3′UTR since these sequences can destabilize a β-globin mRNA and contain A+U-rich elements (AREs) similar to those found in other unstable mRNAs. The presence of a negative element in the HPV-16 early 3′UTR is suggestive because it provides another mechanism by which integration of HPVs in cancer cells leads to the increased expression of the HPV E6 and E7 viral oncogenes. In most HPV-associated cancers, the circular viral genome has integrated into the host genome by linearization within the viral E1 or E2 regions. As a result of disruption of the viral early region, the E6 and E7 genes are expressed as chimeric viral/host mRNAs. Depending on the site of integration in the host genome, the viral early 3′UTR may be replaced by the 3′UTR from an adjacent stable cellular mRNA. The resulting over expression of E6 and E7 would provide a selective growth advantage and clonal expansion of these cells. It should be noted, however, that in their study of the HPV-1 late 3′UTR element, Tan and Schwartz[27] failed to find any negative element in the HPV-16 early 3′UTR. The reason for this discrepancy is unknown, but may be due to use of different cells.

TRANSLATIONAL REGULATION

Papillomavirus gene expression also appears to be regulated at the translational level. *In situ* hybridization and immunoperoxidase studies of an HPV-6-infected vulvar condyloma demonstrated that only a subset of the cells containing L1 mRNA are also expressing high levels of L1 protein, suggesting that L1 expression may be regulated at the translational level.[24] The mechanism for this regulation has not been investigated. However, the BPV-1 L1 mRNA has a noncoding first exon that contains four upstream AUGs which inhibit translation of the L1 protein in transient transfection assays (R. Binder, S. K. Barksdale, A. M. Del Vecchio, J. Tsai and C. C. Baker[9]). Inhibition of translation could be due to either a block to 40S ribosome scanning or to a specific inhibitory peptide(s) translated from one of the uORFs (reviewed in Kozak[19]). *Cis* elements that inhibit translation are also found in the HPV genomes. As mentioned above, the HPV-1 and HPV-16 late 3′UTRs contain such elements.[26,27] In addition, Tan et al.[26] have identified a negative posttranscriptional regulatory element in the L1 ORF that also affects translation of the L1 protein.

CONCLUSIONS

In conclusion, the regulation of papillomavirus gene expression is very complex and takes place at multiple transcriptional and posttranscriptional levels. This review has attempted to show the diversity of posttranscriptional regulatory elements and mechanisms involved in the regulation of papillomavirus gene expression. Additional examples of posttranscriptional regulation in the papillomaviruses have been identified, but could not be included due to space considerations. It is not yet known whether the papillomaviruses share some common posttranscriptional regulatory mechanisms. However, it is interesting to note that although the genomic structure as well as the sequences of many of the viral proteins are highly conserved between the different papillomaviruses, many of the regulatory elements are less well conserved. These differences may reflect the different tissue tropisms of each virus.

REFERENCES

1. Andrews, E. M., and D. DiMaio. "Hierarchy of Polyadenylation Site Usage by Bovine Papillomavirus in Transformed Mouse Cells." *J. Virol.* **67** (1993): 7705–7710.
2. Ashfield, R., P. Enriquez-Harris, and N. J. Proudfoot. "Transcriptional Termination Between the Closely Linked Human Complement Genes C2 and Factor B: Common Termination Factor For C2 and c-Myc?" *EMBO J.* **10** (1991): 4197–4207.
3. Baker, C. C., and J. S. Noe. "Transcriptional Termination Between Bovine Papillomavirus Type 1 (BPV-1) Early and Late Polyadenylation Sites Blocks Late Transcription in BPV-1 Transformed Cells." *J. Virol.* **63** (1989): 3529–3534.
4. Baker, C. C. Unpublished results.
5. Barksdale, S. K., and C. C. Baker. "Differentiation-Specific Expression from the Bovine Papillomavirus Type 1 P2443 and Late Promoters." *J. Virol.* **67** (1993): 5605–5616.
6. Barksdale, S. K., and C. C. Baker. "The Human Immunodeficiency Virus Type 1 Rev Protein and the Rev-Responsive Element Counteract the Effect of an Inhibitory 5′ Splice Site in a 3′ Untranslated Region." *Mol. Cell. Biol.* **15** (1995): 2962–2971.
7. Barksdale, S. K., and C. C. Baker. "Differentiation-Specific Alternative Splicing of Bovine Papillomavirus Late mRNAs." *J. Virol.* **69** (1995): 6553–6556.
8. Berget, S. M. "Exon Recognition in Vertebrate Splicing." *J. Biol. Chem.* **270** (1995): 2411–2414.
9. Binder, R., S. K. Barksdale, A. M. Del Vecchio, J. Tsai, and C. C. Baker. Unpublished results.
10. Burnett, S., J. Moreno-Lopez, and U. Pettersson. "A Novel Spontaneous Mutation of the Bovine Papillomavirus-1 Genome." *Plasmid* **20** (1988): 61–74.
11. Caputi, M., G. Casari, S. Guenzi, R. Tagliabue, A. Sidoli, C. A. Melo, and F. E. Baralle. "A Novel Bipartite Splicing Enhancer Modulates the Differential Processing of the Human Fibronectin EDA Exon." *Nucleic. Acids. Res.* **22** (1994): 1018–1022.
12. Cooke, C., and J. C. Alwine. "The Cap and the 3′ Splice Site Similarly Affect Polyadenylation Efficiency." *Mol. Cell. Biol.* **16** (1996): 2579–2584.
13. Del Gatto, F., M. C. Gesnel, and R. Breathnach. "The Exon Sequence TAGG can Inhibit Splicing." *Nucleic Acids Res.* **24** (1996): 2017–2021.
14. Dietrich-Goetz, W., I. M. Kennedy, B. Levins, M. A. Stanley, and J. B. Clements. "A Cellular 65-kDa Protein Recognizes the Negative Regulatory Element of Human Papillomavirus Late mRNA." *Proc. Natl. Acad. Sci. USA* **94** (1997): 163–168.

15. Enriquez-Harris, P., N. Levitt, D. Briggs, and N. J. Proudfoot. "A Pause Site for RNA Polymerase II Is Associated with Termination of Transcription." *EMBO J.* **10** (1991): 1833–1842.

16. Furth, P. A., W.-T. Choe, J. H. Rex, J. C. Byrne, and C. C. Baker. "Sequences Homologous to 5′ Splice Sites are Required for the Inhibitory Activity of Papillomavirus Late 3′ Untranslated Regions." *Mol. Cell. Biol.* **14** (1994): 5278–5289.

17. Huang, Y. Q., and G. G. Carmichael. "A Suboptimal 5′ Splice Site Is a *Cis*-Acting Determinant of Nuclear Export of Polyomavirus Late mRNAs." *Mol. Cell. Biol.* **16** (1996): 6046–6054.

18. Jeon, S., and P. F. Lambert. "Integration of Human Papillomavirus Type 16 DNA into the Human Genome Leads to Increased Stability of E6 and E7 mRNAs: Implications for Cervical Carcinogenesis." *Proc. Natl. Acad. Sci. USA* **92** (1995): 1654–1658.

19. Kozak, M. "Regulation of Translation in Eukaryotic Systems." *Ann. Rev. Cell Biol.* **8** (1992): 197–225.

20. Manley, J. L., and R. Tacke. "SR Proteins and Splicing Control." *Genes Dev.* **10** (1996): 1569–1579.

21. Proudfoot, N. "Ending the Message Is not so Simple." *Cell* **87** (1996): 779–781.

22. Proudfoot, N. J. "How RNA Polymerase II Terminates Transcription in Higher Eukaryotes." *TIBS* **14** (1989): 105–110.

23. Si, Z. H., B. A. Amendt, and C. M. Stoltzfus. "Splicing Efficiency of Human Immunodeficiency Virus Type 1 Tat RNA Is Determined by Both a Suboptimal 3′ Splice Site and a 10 Nucleotide Exon Splicing Silencer Element Located Within Tat Exon 2." *Nucleic Acids Res.* **25** (1997): 861–867.

24. Stoler, M. H., S. M. Wolinsky, A. Whitbeck, T. R. Broker, and L. T. Chow. "Differentiation-Linked Human Papillomavirus Types 6 and 11 Transcription in Genital Condylomata Revealed by *in Situ* Hybridization with Message-Specific RNA Probes." *Virology* **172** (1989): 331–340.

25. Tacke, R., and J. L. Manley. "The Human Splicing Factors ASF/SF2 and SC35 Possess Distinct, Functionally Significant RNA Binding Specificities." *EMBO J.* **14** (1995): 3540–3551.

26. Tan, W., B. K. Felber, A. S. Zolotukhin, G. N. Pavlakis, and S. Schwartz. "Efficient Expression of the Human Papillomavirus Type 16 L1 Protein in Epithelial Cells by Using Rev and the Rev-Responsive Element of Human Immunodeficiency Virus or the *Cis*-Acting Transactivation Element of Simian Retrovirus Type 1." *J. Virol.* **69** (1995): 5607–5620.

27. Tan, W., and S. Schwartz. "The Rev Protein of Human Immunodeficiency Virus Type 1 Counteracts the Effect of an AU-Rich Negative Element in the Human Papillomavirus Type 1 Late 3′ Untranslated Region." *J. Virol.* **69** (1995): 2932–2945.

28. Wahle, E., and W. Keller. "The Biochemistry of Polyadenylation." *Trends Biochem. Sci.* **21** (1996): 247–250.

29. Zheng, Z. M., P. J. He, and C. C. Baker. "Selection of the Bovine Papillomavirus Type 1 Nucleotide 3225 3′ Splice Site is Regulated Through an Exonic Splicing Enhancer and Its Juxtaposed Exonic Splicing Suppressor." *J. Virol.* **70** (1996): 4691–4699.
30. Zheng, Z. M., P. He, and C. C. Baker. "Structural, Functional, and Protein Binding Analysis of Bovine Papillomarivus Type 1 Exonic Splicing Enhancers." *J. Virology* submitted.

Martijn Huynen,[†][*] **and Danielle Konings**[††]

[†]Santa Fe Institute, 1399 Hyde Park Road, Santa Fe, NM 87501, USA
[*]Biocomputing, EMBL, Meyerhofstrasse 1, 6900 Heidelberg, GERMANY;
E-mail: Martijn.Huynen@EMBL-Heidelberg.de
[‡]Department of Microbiology, Southern Illinois University at Carbondale, Carbondale, Il 62901, USA

Questions About RNA Structures in HIV and HPV

RNA secondary structures play an essential role in human immunodeficiency virus (HIV) and potentially in human papillomavirus (HPV). We give an overview of the known functional secondary structures in HIV-1 and propose a possibly functional secondary structure in HPV-16. Using the examples from HIV-1 and HPV-16, we discuss questions about RNA secondary structure and its evolution: (1) Are there specific features that distinguish functional RNA secondary structures from nonfunctional ones, and can we use these features to search for new functional RNA secondary structures? (2) How does evolution cope with the presence of multiple selective constraints on sequences, in particular, when a sequence simultaneously codes for a protein and for an RNA secondary structure?

INTRODUCTION

We can distinguish various levels of RNA structure. The primary structure is the sequence of the bases (A, G, C, U), whereas the secondary structure is the pattern of base-pair interactions, and the tertiary structure is the conformation of the RNA molecule in three-dimensional space (see Schuster and Stadler, this volume). In the cell, RNA is mainly involved in conveying information from the genes to the cytoplasm (in the case of messenger RNAs), and in expressing that information (e.g., in transfer RNAs and ribosomal RNAs). In RNA viruses, RNA has the additional task of storing and transporting the genetic information for the next generation. Whereas the "readable" information that the RNA carries is in its primary structure, the role of RNA in the expression of that information depends on primary, secondary, and tertiary structure.

Secondary structure can take precedence over primary structure in the sense that for some RNA molecules the secondary structure has been conserved in evolution whereas the primary sequence has not. Well-known examples of this degeneracy are ribosomal RNAs and transfer RNAs (tRNAs), but this has also been observed in functional secondary structures in retroviruses.In these cases, there are many sequences that give rise to the same secondary structure (see Schuster and Stadler, this volume).

Here we will focus on the secondary structures of RNA in human immunodeficiency virus type 1 (HIV-1) and human papillomavirus (HPV). First we give a confined overview of the known functional secondary structures in HIV. Second we present a potentially functional secondary structure in HPV-16. The study of RNA secondary structure and its function in HIV has already been extensively reported and reviewed in the literature. The aim here is to use the known RNA secondary structures in HIV as a starting point for further discussion. For example, we discuss questions concerning constraints on functional RNA secondary structures and how these constraints can be used to design computational techniques for finding functional structures.

AN OVERVIEW OF FUNCTIONAL RNA SECONDARY STRUCTURES IN HIV-1

We will introduce RNA secondary structures of HIV-1 with reference to the role they play during the life cycle of the virus. The organization of the HIV-1 genome, including some of the protein coding regions and the functional RNA secondary structures, is shown in Figure 1.

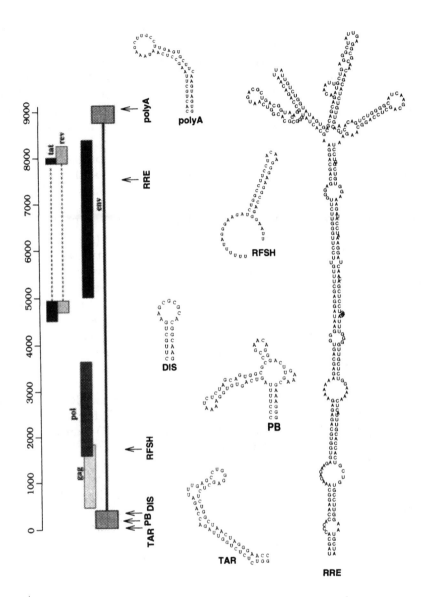

FIGURE 1 An overview of the HIV-1 genome with the positions of the secondary structures that are discussed in the text: transactivation response element (TAR), primer binding site (PB), dimerization initiation site (DIS), ribosomal frameshift hairpin (RFSH), Rev response element (RRE), and polyadenylation site (polyA). The polyA site and the TAR are repeated in the 5′ and 3′ end of the genome, each is shown here only once. In the PB, the nucleotides that bind the primer are in bold. The picture of the RFSH includes the "slippery" heptanucleotide (UUUUU) at its 5′ end (in bold). A number of alternatives have been published for the precise conformation of the RRE. The structure depicted here corresponds largely to the four-hairpin structure published in Mann et al.[21] The genes indicated on the left of the figure are the ones that are mentioned in the text; they are only a subset of all the genes in HIV.

REVERSE TRANSCRIPTION. After the virus enters the cell as a single-stranded RNA molecule, it has to be *reverse transcribed* into DNA. The reverse transcriptase requires a segment of double-stranded RNA to initiate transcription. The RNA molecule, which base-pairs to the single-stranded RNA molecule that will be reverse transcribed, is called the primer. Retroviruses use tRNA molecules as primers. In the case of HIV-1, the primer is usually a $tRNA_3^{lys}$ molecule. It is bound to the primer binding site (PB) in the 5' untranslated region of the viral genome, to which it has a sequence of 18 complementary bases, thus creating the double-stranded RNA to initiate transcription. The PB is located in a conserved secondary structure in which it has been proposed to be largely single-stranded such that the $tRNA_3^{lys}$, for binding to the PB, would not have to compete with other base-pairing interactions.[2] The PB might not be the only RNA secondary structure relevant for reverse transcription. Mutations that alter the secondary structure of TAR, an RNA secondary structure that has been implicated in promoting transcription (see below), drastically reduce levels of reverse transcription.[12] RNA secondary structure might also be involved in regulating the number of mutations during reverse transcription. The introduction of a stem-loop structure in HIV is correlated with a relatively high local mutation rate in reverse transcription.[25] This leads to the fascinating possibility that HIV-1 can locally vary its own mutation rate, e.g., it might have higher mutation rates in parts of proteins that form its primary epitopes and need to keep changing to escape from the immune system than in parts of proteins that have less flexibility to change.

TRANSCRIPTION AND REGULATION OF SPLICING. Two well-known RNA secondary structures are involved in the regulation of expression of HIV after it has been incorporated into the DNA: the transactivation responsive (TAR) element, reviewed in Jones and Peterlin,[15] and the Rev response element (RRE).[9,20,21]

The TAR element resides at the 5' end of the HIV genome, encompassing the first 60 nucleotides of the HIV-1 transcript. The binding of the Tat protein to the TAR RNA structure is required for transcription of the virus. The Tat/TAR interaction is mainly involved in elongation of transcription (see Chao et al., this volume); in the absence of Tat, transcription terminates prematurely. The secondary structure of the TAR is a hairpin with a three-nucleotide bulge at its 5' side and a loop of six nucleotides (Figure 1). The bulge, the loop, and the upper part of the stem are essential for the functioning of TAR. In these regions not only the secondary structure, but also the primary structure is critical for functioning. The Tat protein binds to the bulge, whereas the loop binds a cellular protein. The loop has also been hypothesized to base-pair with a highly conserved complementary sequence, called TAR*, that is located downstream in the *gag* gene.[4] This base-pairing between nucleotides in hairpin loops generates a so-called loop-loop kissing motif. A schematic representation of this interaction is presented in Figure 2.

An interesting aspect of the TAR sequence is that it also plays a role at the DNA level, where it binds cellular proteins as part of the transcriptional promoter.

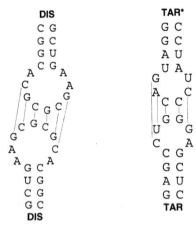

FIGURE 2   Loop-loop kissing motifs in HIV. Nucleotides that form hydrogen bonds (base-pairs) between the loops are connected with lines. Note that the TAR-TAR* motif can occur within a single HIV molecule; the TAR* sequence is located 3′ of the TAR sequence. The DIS-DIS interaction occurs between different HIV molecules and apparently helps initiate their dimerization.

The RRE is located in the *env* gene (Figure 1). It is a relatively large RNA structure (350 nucleotides in HIV-1) that plays a role in the regulation of the transport of unspliced HIV sequences from the cytoplasm. RNA viruses face an interesting dilemma. Their RNA serves both as genetic material and as mRNA that codes for protein. For the expression of most of the proteins, the RNA has to be spliced, making it unfit to serve as genetic material. Early in infection, all of the viral RNA is multiply spliced, producing the mRNAs for, among others, the Tat and Rev proteins. Later in infection, most of the viral RNA leaves the nucleus unspliced or singly spliced. The singly spliced RNA codes for the structural proteins of the viral envelope, whereas the unspliced RNA codes for, among others, the structural Gag protein and the reverse transcriptase. Furthermore, the unspliced RNA is the genetic material for the next generation. In the switch from multiple splicing to single splicing and nonsplicing, the RRE plays a crucial role. The binding of the Rev protein to the RRE promotes the transport of singly spliced and unspliced RNA from the nucleus (see Cullen and also Felber, this volume). Multiple Rev proteins bind to a single RRE, and the binding of the first copy facilitates the binding of other Rev molecules.[21] Once a high enough concentration of Rev protein is reached, the production of genomic material and singly spliced RNA starts.

Note that the Tat/TAR interaction gives rise to a positive feedback loop. Tat stimulates its own production and that of Rev, whereas Rev inhibits the production of both Tat and that of itself. This regulation of viral gene expression with viral proteins is the hallmark of the so-called complex retroviruses.

STRUCTURAL AND POSITIONAL VERSATILITY OF TAR AND RRE. TAR-like secondary structures have been observed in human and simian immunodeficiency viruses, in equine infectious anemia virus (EIAV) and in the bovine immunodeficiency virus (BIV), but not in the lentiviruses caprine arthritis encephalitis virus (CAEV, a goat lentivirus) and visna virus (a sheep lentivirus). Tat binding elements that are functional at the DNA level have been observed in a wider variety of retroviruses, namely CAEV, visna and HTLV-1. The TAR-like secondary structure is always observed at the 5′ end of the viral transcript, although an HIV-1 TAR RNA element that is defective for binding Tat can be complemented by inserting a TAR element > 200 nucleotides downstream of the 5′ end of the virus.[7] The relatively rigid positioning of the TAR element (compared to the RRE, see below) could be explained by the fact that it also has a role at the DNA level as part of the transcriptional promoter; furthermore, its role in the elongation of the transcription process requires positioning near the transcription initiation site. (See Chao et al., this volume.)

The secondary structure of TAR shows considerable variation among the human and simian immunodeficiency viruses. The HIV-2 TAR secondary structure consists of three hairpins enclosed by a helical structure (see Myers and Pavlakis, this volume). The upper two hairpins in HIV-2 TAR result from a duplication of the top part of the HIV-1 TAR hairpin. Interestingly, an HIV-1 variant has been found that has a completely duplicated (and functional) TAR.[10] Cross-reactivity between various Tat proteins and TAR structures is less extensive than documented for the Rev/RRE case (see below). HIV-1 Tat has been shown to transactivate HIV-2 TAR, but transactivation of HIV-2 TAR with HIV-1 Tat is less effective than with HIV-2 Tat. The TAR secondary structures in the other simian lentiviruses closely resemble either HIV-1 TAR or HIV-2 TAR, or take a position intermediate between the two. The distribution of the structural variation in TAR matches the clustering of simian lentiviruses on the basis of their sequences.[1]

RRE-like elements are found in a wider variety of retroviruses than are TAR-like elements: throughout the lentiviruses, and in human T-cell leukemia virus type 1 (HTLV-1) and type 2 (HTLV-2). Furthermore, RNA secondary structure elements that promote transport of unspliced RNA from the nucleus have been found in another group of retroviruses, the relatively simple type D retroviruses such as the Mason-Pfizer monkey virus (MPMV), (see Felber, this volume). In MPMV, the function does not depend on interaction with a viral Rev-like protein, but probably on the interaction with a cellular protein. For an overview of the positions and secondary structures of RRE-like elements, see Figure 3. What is remarkable about the RRE-like structures is their structural and positional versatility. In contrast to secondary structures of ribosomal RNAs, for example, that are strongly conserved in evolution, the secondary structure of RRE-like elements and their positioning within the virus vary strongly. Among HIV-1, HIV-2, and the simian retroviruses, the overall structure of the RRE, containing 5 hairpins enclosed by a long helical structure, is conserved. However,

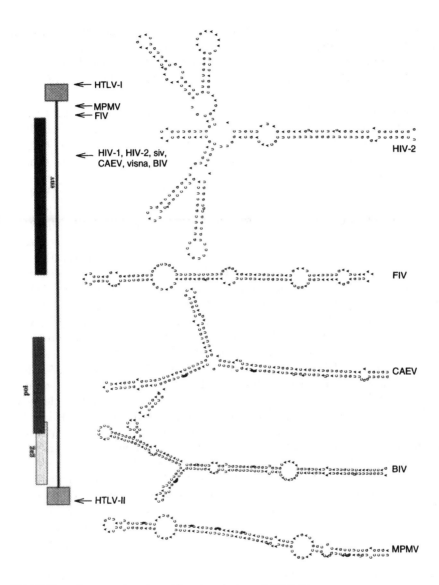

FIGURE 3 Structural and positional variation of RRE-like elements in a variety of retroviruses. The positions are indicated relative to the genomic organization of HIV-1. See the text for detailed information.

some base-pairing patterns are shifted locally: that is to say, homologous nucleotides between the RREs of HIV-1 and HIV-2 occupy nonequivalent positions in the secondary structures.[16] Among HIV and other lentiviruses both the structure and the positions of the RRE-like elements vary. In CAEV and visna virus the RRE-like structure contains three hairpins enclosed by a helical structure that is located in the same region of the *env* gene as the RRE in HIV. As is the case in the RREs of HIV-1 and HIV-2, the base-pairing pattern between the RREs of CAEV and visna has shifted.[26]

To our knowledge, no RRE-like structure has been published for BIV or EIAV. Using computational techniques (see below), we did find a highly significant RNA secondary structure in BIV in the same region of the *env* gene as the RRE in HIV.[14] In the feline immunodeficiency virus (FIV), the RRE-like element consists of a single stem-loop structure that is positioned at the 3′ end of the *env* gene, partly extending into the 3′ untranslated region of the virus. As in FIV, the MPMV RRE-like element consists of a single stem-loop structure that is located completely in the 3′ untranslated region of the virus. Finally, in HTLV-I and HTLV-II, the RRE-like element, in this case named RxRE, is located in the R region, which is present both at the 5′ and at the 3′ end of the genome (note that in HIV-1, the R region contains the TAR element). In HTLV-I the RxRE consists of about 250 nucleotides containing 4 hairpins, and its functional copy is located in the 3′ copy of the R region. In HTLV-II the functional copy of the RxRE is located in the 5′ copy of the R region. Despite this structural and positional variation, cross-reactivity between various Rev-like proteins and RRE-like elements has been widely documented, e.g., between HIV-1 RRE and HTLV-1 Rex (the name of the functional equivalent of Rev in HTLV), between HIV-2 RRE and HIV-1 Rev, HTLV-I Rex or HTLV-II Rex, and between CAEV RRE and HIV-1 Rev or HTLV-1 Rex (see Felber, this volume).

The structural and positional variation of TAR-like and RRE-like elements, combined with the cross-reactivity of Tat/TAR and RRE/Rev are remarkable. Apparently there are relatively few critical constraints on the secondary structure in the TAR and the RRE. As long as some relatively small binding sites are conserved, the rest of the structure serves merely as a scaffold, of which only a general structure needs to be conserved to present the binding sites. In the case of the RRE, the protein binding site is indeed relatively small compared to the overall structure. Since many copies (10–11) of the Rev protein bind to the RRE secondary structure, the cooperativity of the binding probably reduces the constraint of secondary structure compared to constraint at the site where the first protein binds. Note that the presence of cross-reactivity does not imply that all the variation between the RRE-like structures and Rev-like proteins is completely neutral. One expects coevolution between the RNA site and the protein that is supposed to recognize it, and subtle differences in affinity are not alway detected. Furthermore, there are specific examples of non-cross-reactivity, e.g., the RRE of the visna virus (RRE-V) does not cross-react with Rev protein from HIV-1,[27] and the interaction pattern between BIV Tat and TAR shows profound differences with that of the simian Tat/TAR model.[5]

Although the RRE-like elements of closely related viruses (the simian retroviruses, visna, and CAEV) appear to have the same origin, there is not enough nucleic acid sequence similarity among the RRE-like elements across these groups to substantiate a common origin of RRE-like elements in all the retroviruses. The localization of the RRE-like structures of the simian retroviruses, CAEV and visna, and BIV in the same part of the *env* gene (aligned at the amino acid level) does suggest a common origin of their RREs, but might also be the result of evolutionary convergence. The variations between the positioning of these RREs and those of FIV, HTLV-I, and HTLV-II indicates that they have evolved independently. Furthermore, the Rev-like proteins from lentiviruses appear to be unrelated to those of HTLV-I and HTLV-II. The fact that the location of the FIV RRE does not match that of the other lentiviruses is especially striking given that in lentivirus phylogenies that are based on protein sequences FIV is closer to the simian retroviruses than are visna, CAEV, and BIV. This raises the question how easy is it to evolve an RRE-like structure including its Rev-binding domain from scratch? It has been postulated that the Rev-like proteins have arisen from cellular proteins. The fact that the relatively simple type D retroviruses have an RRE-like structure, but do not code for a Rev-like protein and presumably rely on a cellular factor for the Rev function, supports this hypothesis. No cellular proteins with significant sequence similarity to Rev proteins have been found however.

TRANSLATION. The *gag* and *pol* genes in HIV have overlapping reading frames that are out of phase. In order for the ribosome to translate the *pol* gene, it has to shift one position to the 5′ side of the RNA during the translation of the *gag* gene. Such "ribosomal frameshifts" are facilitated by a "slippery" heptanucleotide at the frameshift site and an RNA secondary structure 3′ from the frameshift site. The ribosomal frameshift hairpins are thought to operate by halting the progress of the ribosome as it translates the RNA sequence. This halting can lead to slippage in which the ribosome shifts back one nucleotide to the 5′ end and, hence, ends up translating a different protein. In HIV-1 a significant hairpin structure is present 3′ from the ribosomal frame shift site (Figure 1). In the mouse mammary tumor virus (MMTV) bases in the loop of the frameshift hairpin base-pair with bases 3′ of the hairpin, forming a so-called pseudoknot motif. No direct evidence for such a pseudoknot structure has been found in HIV.

POLYADENYLATION The repeat element (R) that includes the polyadenylation signal "AAUAAA" is present both at the 5′ end and at the 3′ end of the viral genome. The polyadenylation signal has been located, in a wide variety of simian retroviruses, in the loop region of a potential hairpin, (see Berkhout[2] and references therein). It is possible that this secondary structure is related to polyadenylation at the 3′ end, or to another function at the 5′ end. The functional significance of the hairpin structure has been assessed by creating mutations that either stabilize or destabilize it. Both stabilization and destabilization of the hairpin reduce viral replication, pointing to a fine tuning of hairpin stability (see below).

DIMERIZATION AND ENCAPSIDATION. Late in infection two full-length HIV sequences are packed into each virion particle. The two sequences exist as a dimer: they are base-paired to each other at their 5′ end. The temporal order of dimerization and packing is not clear. For both processes, critical sequences and secondary structures have been identified, but results vary depending on the methodology that is used. The sequence that initializes the dimerization of HIV-1, the dimerization initiation signal (DIS), forms a hairpin structure with the sequence "GCGCGC" in the loop, (see Clever,[6] Berkhout,[3] and references therein). Such an autocomplementary sequence can base-pair with its copy in another HIV sequence to start the dimerization, forming a loop-loop kissing motif as has been proposed for the TAR loop structure (see above, and Figure 2). Disruption of the stem structure in which the loop is located prevents dimerization, whereas reconstruction of the stem structure by compensatory mutations reconstitutes dimerization,[6] suggesting that the stem serves to "present" the GCGCGC sequence. Note also that, since the loop sequence is located on top of a stem region, this stem region can be involved in dimerization, in which the 5′ part of the stem in one of the dimerizing sequences base-pairs with the 3′ part of the stem in the other sequence and vice versa. Although variation in the DIS loop sequences has been observed in HIV-1, HIV-2, and simian retroviruses, the loop sequences always contain a six-base palindrome, adding support to the loop-loop kissing model.

Two sequences have been identified that together are essential for packaging the HIV sequences into the virion particles (see McBride and Panganiban[22] and references therein). The sequences form separate hairpin structures. Whereas disruption of either single hairpin only partly reduces encapsidation efficiency (20% to 30%), combined disruption drastically reduces the encapsidation (80%). Restoring the hairpins with compensatory mutations leads to an incomplete recovery of encapsidation efficiency, showing that encapsidation does not only depend on the secondary structure, but might also depend on primary and/or tertiary structure.

A POTENTIALLY FUNCTIONAL RNA SECONDARY STRUCTURE IN HUMAN PAPILLOMAVIRUS 16

HPV is a DNA virus; it does not require RNA secondary structures for packaging or dimerization, since it is packed as DNA. Furthermore, it lacks overlapping reading frames that require a ribosomal frameshift. Hence, one might expect less functional structures in HPV than were found in HIV. There is, however, one aspect of regulation it shares with HIV: the existence of early and late phases of gene expression. The switch from early to late expression involves, as in HIV, a change in the splicing pattern and the export of unspliced RNA from the nucleus.[11] Using computational techniques to find functional secondary structures (described below), we have scanned a set of human papillomaviruses. In contrast to HIV-1, we did not find significant structures that were conserved over most species.

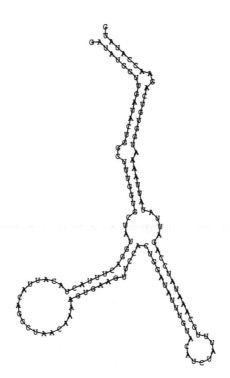

FIGURE 4 The secondary structure of a putative RRE-like element in HPV-16. The structure spans nucleotides 6235 to 6367 in HPV-16. The location of the structure is predicted using computational techniques, (for details see the text).

We did, however, find structures in particular sequences that could be considered significant. As a striking example, a large structure was found in the L1 gene of HPV-16 (Figure 4). Its large size and location within a gene that is only expressed in the late phase of gene expression suggests that it could potentially function as an RRE analog in HPV-16: i.e., as an element that promotes the export of unspliced RNA from the nucleus.

WHAT ARE THE PROPERTIES OF FUNCTIONAL RNA SECONDARY STRUCTURES?

Although all RNA molecules have a potential secondary structure, that in itself does not imply that the secondary structure serves a particular biological function. At one extreme, the function of the RNA can be solely to encode a protein which is only determined by the primary structure of the RNA. We must ask, "Are there properties of functional RNA secondary structures that distinguish them from non-functional secondary structures, and can we use these properties to find new functional structures?" Given the above-described structural versatility of structures with similar functions, searching for functional secondary structures by focusing on specific structural motifs is not a good strategy. However, features have been described that do distinguish functional secondary structures from secondary structures of random sequences. We describe two features of functional RNA secondary structures and how they can be used to find new functional structures. In order to do this, we first give a short explanation of the thermodynamic model of RNA secondary structure prediction.

THE THERMODYNAMIC MODEL OF RNA SECONDARY STRUCTURE PREDICTION

The thermodynamic model of RNA secondary structure prediction relies on the assumption that free-energy of RNA secondary structure formation can be approximated by adding up the experimentally determined free energies of its elements (base-pairs, hairpin loops, bulges etc.). The secondary structure with the lowest free energies which can be found using a dynamic programming algorithm has, in thermodynamic equilibrium, the highest likelihood of occurring.[28] This approach, in combination with experimental methods of secondary structure analysis, has become a major tool in research on RNA structure. But secondary structure prediction by free energy minimization faces the problem that for any sequence there is an exponentially large number of alternative structures. For a sequence of length N, there are about 1.8^N possible structures (see Schuster and Stadler, this volume). While in thermodynamic equilibrium, the structure with the lowest free energy has the highest probability; this probability can become very small for long sequences due to the high number of alternative structures. For example, the probability of the lowest free-energy structure for a random sequence of the length of a ribosomal RNA (1500 nucleotides) is generally smaller than 10^{-45}. A more interesting quantity than the probability of a specific large structure is the probability of small substructures. RNA secondary structure prediction is inherently nonlocal and the probability of any substructure within the sequence depends on the entire sequence. This approach, which has been formalized by McCaskill,[23] focuses on the smallest substructure, the base-pair. An algorithm calculates the comprehensive base-pair probability distribution based on the free energies and resulting probabilities of all structures. A graphical representation of such a base-pair probability matrix is shown in Figure 5.

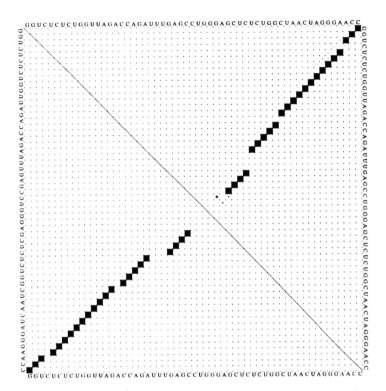

FIGURE 5   A graphical representation of a base-pair probability matrix, in this case of the TAR element of HIV-1. The upper-right triangle contains the base-pairing probability matrix. The size of the squares is proportional to the pairing probability. The lower-left triangle displays the minimum free-energy structure for comparison. Hairpin loops appear as diagonal patterns close to the separating line between the two triangles, with the distance from this line indicating the loop size. Internal loops and bulges appear as shifts and gaps in the diagonal patterns.

"WELL DEFINEDNESS" OF RNA SECONDARY STRUCTURE

In order to analyze the information in the base-pairing probability matrix for long sequences, we need to condense it. An interesting property of the base-pairing probability matrix is its "well-definedness": i.e., the degree to which the base-pair probability distribution per base is dominated by a single base-pair interaction, or by the absence of base-pairing (e.g., if the base is generally located in a loop region).

We express the extent to which the base-pairing probability distribution is dominated by a single base-pairing interaction, or by the absence of base-pairing, with the Shannon-entropy (S) where

$$S_i = -\sum_j P_{i,j} \log P_{i,j}\,,$$

$P_{i,j}$ is the probability of base-pairing between bases i and j, and $P_{i,j=i}$ is the probability that the base i does not pair with any other base. The lower the S value, the stronger the distribution is dominated by a single or a few base-pairing probabilities. In other words, layer S values reflect the uncertainty we have about the base-pairing.

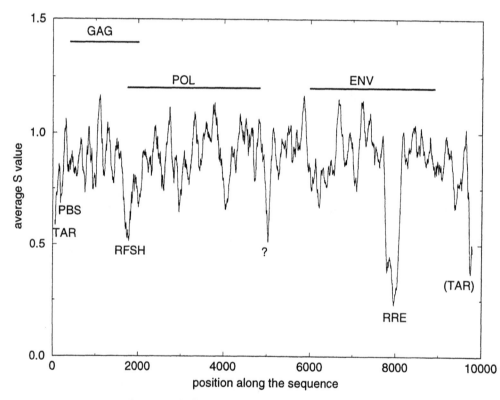

FIGURE 6 "Well-definedness" of the base-pairing probability distribution for 21 aligned HIV-1 sequences. The well-definedness of the base-pairing is measured as the Shannon-entropy of the base-pairing probability distribution per base. Minima correspond to secondary structures with relatively few alternatives (see text). The lowest minima in the figure correspond to known functional RNA secondary structures in HIV. The abbreviations of the secondary structures are as in Figure 1. One minimum, indicated with a question mark, does not correspond to a known functional structure. The positions of the HIV *gag*, *pol*, and *env* genes are indicated for comparison.

We plotted the average S score per position for 21 aligned HIV-1 sequences (Figure 6). The results show that most of the minima in the S scores correspond to known functional secondary structures. Hence, the problem that was sketched earlier for secondary structure prediction—the presence of a huge number of alternative secondary structures—is one that not only the researcher, but also the species, faces. Nevertheless, adaptation has resulted in sequences with relatively well-defined secondary structures. This exemplifies the flexibility of the RNA sequence to RNA secondary structure mapping. For any structure there are many sequences, but not all those sequences give rise to that secondary structure with the same probability. Evolution chooses which sequence to use depending on the requirements on the thermodynamic stability of the structure, . In general it appears to choose sequences with a relatively well-defined secondary structure. Ribosomal RNAs of prokaryotic species that live at temperatures above 80° have been shown to have better defined secondary structures than the ribosomal RNAs of species that live at lower temperatures.[13] Thus well-definedness seems to be adapted to the environment. In Figure 6 there are a few minima that do not correspond to known functional RNA secondary structures. These might be noise, but could also be new, as yet undiscovered structures.

SEARCHING FOR FUNCTIONAL RNA SECONDARY STRUCTURES BY RELATIVELY LOW FREE ENERGY

A second method for finding functional secondary structures is based on a related principle. It searches for regions with unusual folding properties according to two principles: (1) The free energy (e) is significantly lower than the mean free energy of random sequences with the same nucleotide frequencies and length (er) [expressed as significance score = (e-er)/sdr, where sdr is the standard deviation of the random sample]. (2) The free energy (e) is relatively low compared to the mean of the free energies of all other segments of the same length in the biological sequence (eb) [expressed as stability score = (e-eb)/sdr, where sdr is the standard deviation of the sample]. Segments are of particular interest when both scores are significantly low.[18] The analysis is usually performed with different window sizes (e.g., ranging from 50–200 nucleotides). This method has been very powerful in the detection of a wide range of functional RNA secondary structures, in particular in viruses, (e.g., Konings et al.[17] and Lee and Zuker[19]).

Functional RNA secondary structures seem to have a relatively low free energy and are relatively well defined. In Figure 7 we show a plot of the HPV-16 late transcript for both methods. Both methods show a strong minimum in the area 6200–6400. The minimum free-energy structure of that region is proposed in Figure 4. This structure is currently under investigation.

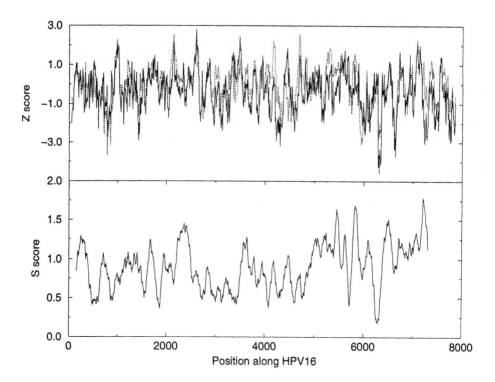

FIGURE 7 Significance and "well-definedness" scores for HPV-16. The top part of the figure shows the significance of the free energy of segments of 100 nucleotides. The significance is calculated relative to random sequences with the same base content (continuous line) and relative to the rest of the sequence (dotted line). The bottom part of the figure shows the S score. All three scores show a minimum around positions 6200–6400.

SELECTION AGAINST HIGH STABILITY

Although in the above illustrations we have put emphasis on the importance of well-defined secondary structures with a relatively low free energy, one should keep in mind that are excessively stable structure might be detrimental to the "fitness" of a sequence. Translation and reverse transcription have to melt RNA secondary structures along their path, and increased stability of secondary structure reduces the rate at which they progress along the RNA (sometimes this property is actually used, e.g., in the case of the ribosomal frameshift hairpin; see above). As previously noted, an increased stability of the polyadenylation hairpin leads to reduced levels of replication. In HIV-1, the two lowest bulges in TAR that have not been implicated in any specific function might serve to simply decrease the stability of the secondary structure. An informative example of selection against an excessively stable structure has been demonstrated in the bacteriophage MS2 (an RNA virus that parasitizes *Escherichia coli*). Mutations that increase the stability of a certain hairpin at the 5′ end of the coat protein gene reduce the rate of translation. Furthermore, in evolution experiments that were initiated with a hairpin with an increased or a decreased stability (relative to the wild type), mutations were selected that returned the secondary structure of the hairpin to its original stability.[24] Such selection against an overly high stability might explain why not all secondary structures that are known to be functional in the 5′ leader sequence of HIV-1 are well defined (Figure 6).

HOW DOES EVOLUTION COMPROMISE BETWEEN THE DIFFERENT CONSTRAINTS ON A MOLECULE?

Some of the secondary structures that are known to be functional in HIV are located in protein coding regions. The RNA sequences in these regions code for a protein as well as an RNA secondary structure. Striking examples of such "multiple coding" are ribosomal frameshift hairpins (see above). The sequences encompassing the frameshift hairpins are actually triply coding—for two amino acid sequences and one RNA structure. In HIV such a structure is located downstream of the frameshift between the *gag* and the *pol* reading frames. Another example of multiple coding in HIV is the RRE structure. Given the length of the functional RRE structure in HIV-1 (around 350 nucleotides) this overlap of codes is rather unique. Multiple coding is not necessarily constrained to protein and RNA structure. DNA structure might also play a role. The TAR sequence (Figure 1) that is located at the 5′ end of the HIV virion has both a functional RNA secondary structure and a functional DNA structure. It is, in general, hard to distinguish between the different functions of the RNA structures in the 5′ end of the HIV genome. For example, one may question to what extent we can discriminate between the encapsidation and dimerization processes. Mutations that destroy the DIS site lead to a reduction in viral encapsidation.[3] Furthermore, the TAR structure, besides its role in promoting transcription, seems to have an additional role in reverse transcription.[12]

The phenomenon of multiple coding raises several interesting questions: which evolutionary pressures push a genome toward multiple coding and how does evolution compromise between constraints at different levels, such as DNA, RNA, and protein structures? Some factors that may be involved in the evolutionary pressure are: (i) The limited genome size as a result of relatively low accuracy of replication as observed in viruses, a phenomenon known as the "information threshold."[8] Multiple coding, such as RNA/protein overlap and gene sharing in general, compromises with this pressure.[16] (ii) The regulation and interdependence of different functions, for example, the involvement of the RNA hairpins in ribosomal frameshifting.

To understand how these constraints are met by organisms, one approach to this problem would be to investigate the redundancy in sequence-to-structure mappings, i.e., these are often many-to-one mappings: for every desired structure there are many alternative sequences. The best-known example of this is the redundancy in the genetic code. There are 64 codons for 20 amino acids and stop codons. Another redundancy that could be investigated is in the mapping from RNA sequence to RNA secondary structure. Many sequences have the same minimum free-energy secondary structure (see Schuster and Stadler, this volume). Also, structural flexibility, particularly observed in the fast evolving viral systems, allows for quite some playground in evolution. Relatively large parts of RNA and protein structures act as rough scaffolds, do not express specific structural and sequence information, and thus allow for the compromise of different codes. It is, in this light, an interesting question to what extent the structural versatility of the RRE is a result of the fact that it lies in a protein coding region. What evolution has really played with goes far beyond a simple story and a common set of factors. With time and the study of more biological systems, our understanding will grow and evolutionary relationships should become more clear.

ACKNOWLEDGMENTS

Martijn Huynen thanks Ian Mattaj for interesting discussions. Part of this work was done under the auspices of the US Department of Energy and supported by the Santa Fe Institute.

REFERENCES

1. Berkhout, B. "Structural Features in TAR RNA of Human and Simian Immunodeficiency Viruses: A Phylogenetic Analysis." *Nucl. Acids Res.* **20** (1992): 27–31.

2. Berkhout, B. "Structure and Function of the Human Immunodeficiency Virus Leader RNA." *Prog. Nucleic Acid Res. Mol. Biol.* **54** (1996): 1–34.

3. Berkhout, B., and J. L. Wamel. "Role of the DIS Hairpin in Replication of Human Immunodeficiency Virus Type 1." *J. Virol.* **70** (1996): 6723–6732.

4. Chang, K., and I. Tinoco, Jr. "Characterization of a 'Kissing' Hairpin Complex Derived from the HIV Genome." *Proc. Natl. Acad. Sci., USA* **91** (1994): 8705–8709.

5. Chen, L., and A. D. Frankel. "An RNA-Binding Peptide from Bovine Immunodeficiency Virus Tat Protein Recognizes an Unusual RNA Structure." *Biochemistry* **33** (1994): 2708–2715.

6. Clever, J. L., M. L. Wong, and T. G. Parslow. "Requirements for Kissing-Loop-Mediated Dimerization of Human Immunodeficiency Virus RNA." *J. Virol.* **70** (1996): 5902–5908.

7. Churner, M. J., A. D. Lowe, M. J. Gait, and J. Karn. "The RNA Element Encoded by the Trans-Activation-Responsive Region of Human Immunodeficiency Virus Type 1 is Functional when Displaced Downstream of the Start of Transcription." *Proc. Natl. Acad. Sci. USA* **92** (1995): 2408–2412.

8. Eigen, M., and P. Schuster. *The Hypercycle.* Berlin: Springer-Verlag, 1979.

9. Elahem, D., D. M. Powell, and A. I. Dayton. "Functional Analysis of CAR, the Target Sequence for the Rev Protein of HIV-1." *Science* **246** (1989): 1625–1629.

10. Emiliani, S., C. Delsert, C. David, and C. Devaux. "The Long Terminal Repeat of the Human Immunodeficiency Virus Type 1 GER Isolate Shows a Duplication of the TAR Region." *AIDS Res. Hum. Retroviruses* **10** (1994): 1751–1752.

11. Furth, P. A., W. Choe, J. H. Rex, J. C. Byrne, and C. C. Baker. "Sequences Homologous to 5′ Splice Sites Are Required for the Inhibitory Activity of Papillomavirus Late 3′ Untranslated Regions." *Mol. Cell. Biol.* **14** (1994): 5278–5289.

12. Harrich, D., C. Ulich, and R. B. Gaynor. "A Critical Role for the TAR Element in Promoting Efficient Human Immunodeficiency Virus Type 1 Reverse Transcription. " *J. Virol.* **70** (1996): 4017–4027.

13. Huynen, M. A., R. Gutell, and D. A. M. Konings. "Assessing the Reliability of RNA Folding Using Statistical Mechanics." *J. Mol. Biol.* **267** (1997): 1106–1112.

14. Huynen, M. A. Unpublished results.

15. Jones, K. A., and B. M. Peterlin. "Control of RNA Initiation and Elongation at the HIV-1 Promoter." *Ann. Rev. Biochem.* **63** (1994): 717–743.

16. Konings, D. A. M. Coexistence of Multiple Codes in Messenger RNA Molecules." *Comp. & Chem.* **16** (1992): 153–163.

17. Konings, D. A. M., M. A. Nash, J. V. Maizel, and R. B. Arlinghaus. "Novel GACG-Hairpin Pair Motif in the 5′ Untranslated Region of Type C Retroviruses Related to Murine Leukemia Virus." *J. Virol.* **66** (1992): 632–640.

18. Le, S.-Y., J.-H. Chen, and J. V. Maizel. "Thermodynamic Stability and Statistical Significance of Potential Stem-Loop Structures Situated at the Frameshift Sites of Retroviruses." *Nucl. Acids Res.* **17** (1989): 6143–6152.

19. Le, S.-Y., and M. Zuker. "Common Structures of the 5′ Non-Coding RNA in Enteroviruses and Rhinoviruses. Thermodynamical Stability and Statistical Significance. " *J. Mol. Biol.* **216** (1990): 729–741.

20. Malim, M. H., J. Hauber, S.-Y. Le, J. V. Maizel, and B. R. Cullen. "The HIV-1 Rev Trans-Activator Acts Through a Structured Target Sequence to Activate Nuclear Export of Unspliced Viral mRNA." *Nature* **338** (1989): 254–257.

21. Mann, D. A., I. Mikaelian, R. W. Zemmel, S. M. Green, A. D. Lowe, T. Kimura, M. Singh, P. J. Butler, M. J. Gait, and J. Karn. "A Molecular Rheostat. Co-operative Rev Binding to Stem I of the Rev-Response Element Modulates Human Immunodeficiency Virus Type-1 Late Gene Expression." *J. Mol. Biol.* **241** (1994): 193–207.

22. McBride, M. S., and A. T. Panganiban. "The Human Immunodeficiency Virus Type 1 Encapsidation Site is a Multipartite RNA Element Composed of Functional Hairpin Structures." *J. Virol.* **70** (1996): 2963–2973.

23. McCaskill, J. S. "The Equilibrium Partition Function and Base PAir Binding Probabilities for RNA Secondary Structure Biopolymers." **29** (1990): 1105–1119.

24. Olsthoorn, R. C., N. Licis, and J. van Duin. "Leeway and Constraints in the Forced Evolution of a Regulatory RNA Helix." *EMBO J.* **13** (1994): 2660–2668.

25. Pathak, V. K., and H. M. Temin. "5-Azacytidine and RNA Secondary Structure Increase the Retrovirus Mutation Rate." *J. Virol.* **66** (1992): 3093–3100.

26. Saltarelli, M., G. Querat, D. A. M. Konings, R. Vigne, and J. E. Clements. "Nucleotide Sequence and Transcriptional Analysis of Molecular Clones of HIV which Generate Infectious Virus." *Virology* **179(3)** (1988): 387–393.

27. Tiley, L. S., and B. R. Cullen. "Structural and Functional Analysis of the Visna Virus Rev-Response Element." *J Virol.* **66** (1992): 3609–3615.

28. Zuker, M., and P. Stiegler. "Optimal Computer Folding of Larger RNA Sequences Using Thermodynamics and Auxiliary Information." *Nucl. Acids. Res.* **9** (1981): 133–148.

Peter Schuster[†‡]* and Peter F. Stadler[†‡]
[†] Institut für Theoretische Chemie, Universität Wien, Währingerstraße 17, A-1090 Wien, AUSTRIA; *E-mail: pks@tbi.univie.ac.at
[‡] The Santa Fe Institute, Santa Fe, NM 87501; *E-mail: pks@santafe.edu

Sequence Redundancy in Biopolymers: A Study on RNA and Protein Structures

Mapping sequences onto biopolymer structures is characterized by redundancy since the number of sequences exceeds the number of structures. The degree of redundancy depends on the notion of structure. Two classes of biopolymers, RNA molecules and proteins, are considered here in detail. A general feature of sequence-to-structure mappings is the existence of a few common and many rare structures. Consequences of redundancy and frequency distribution of RNA structures are *shape-space covering* and the existence of extended *neutral networks*. Populations migrate on neutral networks by a diffusion-like mechanism. Neutral networks are of fundamental importance for evolutionary optimization since they enable populations to escape from local optima of fitness landscapes.

1. THE ORIGIN OF REDUNDANCY IN BIOPOLYMER STRUCTURES

The notion of selective neutrality appears early on in Charles Darwin's *Origin of Species*. Until now, nevertheless, no quantitatively satisfactory concept of sequence redundancy has been given in evolutionary theory. Motoo Kimura's neutral evolution[20] is based on neutrality in the strict sense of identical fitness values. Tomoko Ohta[26] extended Kimura's concept to variable environments. In the generic case it is very unlikely that a fittest genotype is optimal under all conditions. Commonly, genotypes will be slightly deleterious in most of the environments and best adapted only under very few circumstances. Ohta's theory, often addressed as "near neutral theory," makes an important contribution to the understanding of evolution: neutrality with respect to selection does not mean fitnesses that are identical to the very last digit, rather it implies a band of fitness values among which selection is unable to distinguish. Whether or not two or more genotypes produce phenotypes of indistinguishable fitness, which are therefore neutral, is not only a matter of properties in isolation but also a result of environmental conditions, selection constraints, and population size.

Mapping of genotypes into fitness values is a core issue of evolutionary biology; this mapping is commonly simplified by partitioning in two steps:

$$\textbf{Genotype} \quad \Longrightarrow \quad \textbf{Phenotype} \quad \Longrightarrow \quad \textbf{Fitness}\ .$$

Exceptions to this scheme are some model landscapes that assign fitness values to genotypes more or less randomly, for example, spin glass models or the closely related *N-k* model of Stuart Kauffman.[19] Genotype-phenotype mappings are generally too complicated to be analyzed by rigorous techniques, although *in vitro* evolution of molecules reduces this map to relations between polynucleotide sequences and biopolymer structures and functions. These relations are also a primary subject of molecular biophysics. Here we shall review the first step of this combined map. It deals with the formation of biomolecular structures through folding of genotypes, including RNA or DNA sequences of polynucleotides as well as amino acid sequences of proteins. Neutrality of structures implies redundancy of sequences in the sense that two or more sequences give rise to the same structure.

In the precise view of X-ray crystallography of biopolymers, sequence redundancy is nonexistent: Small as they may be there are always differences in atomic coordinates that make structures unique. The crystallographic notion of structure, however, is vastly different from biochemical and evolutionary intuitions, hence, protein and RNA structures can often be represented by wire diagrams. Phylogenetic conservation of structure is discussed, for example, by comparison of backbone foldings. High precision, nevertheless, is required for active sites of enzymes and ribozymes as well as for specific recognition sites of regulators' biopolymers. Conserved positions and residues are few compared to a hundred and more monomer units that can be changed without substantially altering biopolymer function. Whether or not a change caused by a substitution is significant is not a matter of structure but also a question of environmental fluctuations and selection constraints.

Near-neutral theory defines a band of almost optimal fitness values and, thus, variations in structure and properties are tolerated. This fact calls for a coarse-grained notion of structure. An operational coarse graining that is suitable for modeling evolution is not available yet. There are, however, established notions of structural coarse graining that can be analyzed in detail. In sections 2 and 3 we shall discuss coarse-grained sequence-structure relations of RNA molecules and proteins. The consequences of redundancy in sequence structure mappings for the course of evolutionary adaptations are reviewed in section 4. The concluding section, section 5, compares sequence redundancy in the two classes of biopolymers.

2. THE SEQUENCE STRUCTURE MAP OF RNA

Mapping RNA genotypes onto phenotypes becomes accessible to straightforward analysis when the phenotype can be identified with the molecular structure of the RNA molecules. This is the case with *in vitro* evolution of RNA and, presumably, also with simple RNA viruses whose life cycles are determined by the structure of the viral RNA. Molecular structures of RNA molecules are complicated objects. Often they cannot be represented by only a single conformation, in which case a statistical description by means of the matrix of base pairing probabilities[24] is appropriate. We shall not discuss this issue here, but instead assume that the minimum free-energy conformation is representative for the properties of the molecule.

2.1 RNA SECONDARY STRUCTURES

Mapping RNA genotypes into phenotypes requires a solution to the structure prediction problem.[32] Current knowledge of three-dimensional structures of RNA molecules, however, is rather limited: only very few structures have been determined by crystallography and NMR spectroscopy. Needless to say, spatial structures of RNA molecules are also very hard to predict by computations based on minimization of potential energies and molecular dynamics simulations. The so-called secondary structure of RNA is a coarse-grained version of structure that lists Watson-Crick (**GC** and **AU**) and **GU** base pairs. A secondary structure can be

represented by a planar graph without knots or pseudoknots.[1] Secondary structures are conceptionally much simpler than three-dimensional structures and allow
rigorous mathematical analysis[36] as well as large scale computations by means
of algorithms based on dynamic programming[37] and implementation on parallel
processors.[14] RNA secondary structure predictions are more reliable than those of
full spatial structures. In addition, the definition of RNA secondary structures allows formally consistent distance measures (D) in shape space.[10] Some statistical
properties of RNA secondary structures have been shown to depend very little on
choices of algorithms and parameter sets.[35]

RNA secondary structures provide an excellent model system for the study
of global relations between genotypes and phenotypes. The conventional approach
of structural biology determining structures for single sequences is extended to
a general concept that considers sequence structure relations as (noninvertible)
mappings from sequence space into shape space.[10,28,30]

2.2 COMMON AND RARE RNA STRUCTURES

Application of combinatorics to RNA secondary structures[36] allows derivation of
an asymptotic expression from a simple recursion for the number of acceptable
structures that can be formed by sequences of chain length n[29]

$$S_n \approx 1.4848 \times n^{3/2} (1.8488)^n . \tag{1}$$

Equation (1) is based on two assumptions: (i) the minimum stack length is two
base pairs ($n_{\text{stack}} \geq 2$, i.e., isolated base pairs are excluded) and (ii) the minimal
size of hairpin loops is three ($n_{\text{loops}} \geq 3$). The number of sequences is given by 4^n
for natural RNA molecules and by 2^n for **GC**-only or **AU**-only sequences. For both
classes of RNA molecules there are more sequences than secondary structures. In
the evolutionary relevant cases there will be a large number of sequences folding into
the same secondary structure ψ. We call the set $S(\psi)$ of sequences folding into ψ the
neutral set of structure ψ. For natural RNA molecules this general picture does not
change qualitatively if nonnested pseudoknots are allowed as well.[13] The number of
acceptable structures with pseudoknots increases asymptotically with $S_n \propto 2.35^n$
instead of 1.85^n, as was derived for the structures without pseudoknots (see Eq.
(1)). The result was derived without applying a constraint for the stereochemistry
of pseudoknots and thus represents an upper limit for the number of acceptable
structures. It shows, however, that the view of RNA shape space derived from
secondary structures does not change drastically when tertiary interactions are
included.

---

[1] The precise definition for an acceptable secondary structure is: (i) base pairs are not allowed between neighbors in the sequences $(i, i + 1)$ and (ii) if (i, j) and (k, ℓ) are two base pairs then (apart
from permutations) only two arrangements along the sequence are acceptable: $(i < j < k < \ell)$ and
$(i < k < \ell < j)$, respectively.

TABLE 1 Common secondary structures of **GC**-only sequences.

	# Sequences		# Struct.		**GC**[1]	
n	4^n	2^n	S_n	$\tilde{S}_{\mathbf{GC}}$	R_c	n_c
7	16,384	128	6	2	1	120
10	1.05×10^6	1,024	22	11	4	859
15	1.07×10^9	32,768	258	116	43	28,935
20	1.10×10^{12}	1.05×10^6	3,613	1,610	286	902,918
25	1.13×10^{15}	3.36×10^7	55,848	18,590	2,869	30,745,861
30	1.15×10^{18}	1.07×10^9	917,665	218,820	22,718	999,508,805

[1] The total number of minumum free-energy secondary structures formed by **GC**-only sequences is denoted by $\tilde{S}_{\mathbf{GC}}$; R_c is the rank of the least frequent common structure and, thus, is tantamount to the number of common structures; and n_c is the number of sequences folding into common structures.

Not all acceptable secondary structures are actually formed as minimum free-energy structures. The number of stable secondary structures, $\tilde{S}_n$, was determined by exhaustive folding[11,12] of all **GC**-only sequences with chain lengths up to $n = 30$ (Table 1). The fraction of acceptable structures obtained as minimum free-energy structures through folding **GC** sequences is between 20% and 50%. This fraction is decreasing with increasing chain length n. The best estimate of the exponential increase of minimum free-energy structures from our data is $\tilde{S}_n \propto 1.65^n$.

Secondary structures are properly grouped into two classes, common ones and rare ones. A straightforward definition of common structures is found to be very useful: A structure ψ is *common* if it is formed by more sequences than the average. That is to say,

$$|S(\psi)| \geq \overline{|S|} = \frac{\kappa^n}{\tilde{S}_n}, \qquad (2)$$

where κ denotes the size of the alphabet ($\kappa = 2$ for **GC**-only or **AU**-only sequences and $\kappa = 4$ for natural RNA molecules).

The results of exhaustive folding suggest two important general properties of this definition of common structures[11,12]: (i) the common structures represent only a small fraction of all structures, and this fraction decreases with increasing chain length; and (ii) the fraction of sequences folding into the common structures increases with chain length and approaches unity in the limit of long chains. Thus, for sufficiently long chains almost all RNA sequences fold into a small fraction of the secondary structures. The effective ratio of sequences to structures is larger than computed from Eq. (2) since only common structures play a role in natural evolution and in evolutionary biotechnology.

2.3 THE TOPOLOGY OF NEUTRAL SETS

The shape or topology of neutral sets has important implications for the evolution of both nucleic acids and proteins and for *de novo* design. For example, it has been frequently observed that seemingly unrelated protein sequences have essentially the same fold, (see Murzin[25] and the references therein). Similarly, the genomic sequences of closely related RNA viruses show a large degree of sequence variation while sharing many conserved features in their secondary structures.[15,27] Whether these may have originated from a common ancestor, or whether they must be the result of convergent evolution, depends on the geometry of the neutral sets $S(\psi)$ in sequence space. Another well-known example is represented by the clover leaf secondary structure of tRNAs: The sequences of different tRNAs have very little base similarity[7] but nevertheless fold into the same secondary-structure motif.

Inverse folding can be used to determine the sequences that fold into a given structure. For RNA secondary structures an efficient inverse-folding algorithm is available.[14] It was used to show that sequences folding into the same structure are (almost) randomly distributed in sequence space. It was also noticed in early work on RNA secondary structures[9] that a substantial fraction of point mutants are neutral in the sense that the corresponding sequences fold into the same secondary structure. Detailed data can be found in Grüner et al.[11]

Two approaches have been applied so far to study the topology of neutral sets: a mathematical model of genotype-phenotype mapping based on random graph theory[28] and exhaustive folding of all sequences with a given chain length n.[12] The mathematical model assumes that sequences forming the same structure are distributed randomly (in the space of compatible sequence, see below) and it uses the fraction λ of neutral neighbors as the only input parameter. If λ is large enough, this model makes two rather surprising predictions:

1. There is *shape-space covering*, that is, in a moderate size ball centered at any position in sequence space there is a sequence x that folds into any prescribed secondary structure ψ.
2. The neutral sets $S(\psi)$ of all common structures form networks that percolate sequence space.

In the following two sections we shall discuss these predictions in more detail for the RNA case.

2.4 SHAPE-SPACE COVERING

It is relatively straightforward to compute a spherical environment (around any randomly chosen reference point in sequence space) that contains at least one sequence (on the average) for every common structure. The radius of such a sphere, called the covering radius r_{cov}, can be estimated from simple probability arguments[31]:

$$r_{cov} = \min \left\{ h \mid B_h \geq \tilde{S}_n \right\}, \tag{3}$$

with B_h being the number of sequences contained in a ball of radius h. The covering radius is much smaller than the radius of sequence space. The covering sphere represents only a small connected subset of all sequences but contains, nevertheless, all common structures and forms an evolutionarily representative part of shape space.

Numerical values of covering radii are presented in Table 2. In the case of natural sequences of chain length $n = 100$ a covering radius of $r_{cov} = 15$ implies that the number of sequences that have to be searched in order to find all common structures is about 4×10^{24}. Although 10^{24} is a very large number (exceeds the

TABLE 2 Shape-space covering radius for common secondary structures.

n	Covering Radius r_{cov}				$B_{r_{cov}} / 4^{\kappa 2}$
	Exhaustive Folding		Estimate from Eq.(3)[1]		
	GC^3	AU	$\kappa = 2$	$\kappa = 4$	
20	3 (3.4)	2	4	2	3.29×10^{-9}
25	4 (4.7)	2	4	3	4.96×10^{-11}
30	6 (6.1)	3	7	4	7.96×10^{-13}
50	10.7^4	7^4	12	6	7.32×10^{-20}
70	15.6^4	11.5^4	18	10	8.75×10^{-27}
100	22.9^4	17.3^4	26	15	4.52×10^{-37}

[1] The covering radius is estimated by means of a straightforward statistical estimate based on the assumption that sequences folding into the same structure are randomly distributed in sequence space. Eq. (1) is used as an estimate for $\tilde{S}_n$.

[2] Fraction of **AUGC** sequence space that has to be searched on the average in order to find at least one sequence for every common structure.

[3] Exact values derived from exhaustive folding are given in parantheses.

[4] Upper bounds from Grüner et al.,[12] Table 5.

capacities of all currently available polynucleotide libraries), it is negligibly small compared to the size of the entire sequence space that contains 1.6×10^{60} sequences. Exhaustive folding allows testing of the estimates derived from simple statistics.[12] The agreement for **GC**-only sequences of short chain lengths is surprisingly good. The covering radius increases linearly with chain length that has a slope around $1/4$. The fraction of sequence space that is required to cover shape space thus decreases exponentially with increasing size of RNA molecules (Table 2). We remark that, nevertheless, the absolute numbers of sequences contained in the covering sphere increase (also exponentially) with the chain length.

2.5 NEUTRAL NETWORKS

Every common structure is formed by a large number of sequences and, hence, it is highly important to know how sequences folding into the same structure are organized in sequence space. In order to understand the geometric structure of a single neutral set $S(\psi)$, we shall need the *space of compatible sequences* of a secondary structure ψ. A sequence is *compatible* with a structure when it can, in principle, fold into this structure. The sequence requires complementary bases in all pairs of positions forming a base pair in the structure. When a sequence is compatible with a structure, then the latter is necessarily among the minimum free-energy or suboptimal foldings of the RNA molecule; a compatible sequence x might, but need not, form the structure ψ under minimum free-energy conditions.

The neutral network of a structure is the subset of its compatible sequences that actually form the structure under the minimum free-energy conditions. Following a strictly mathematical approach[28] neutral networks are modeled by random graphs in sequence space. The analysis is simplified through partitioning of sequence space into a subspace of unpaired bases and a subspace of base pairs. Neutral neighbors in both subspaces are chosen at random and connected to yield the edges of the random graph that is representative for the neutral network. The parameter λ measures the mean fraction of neutral neighbors in sequence space, therefore the statistics of random graphs are studied as a function of λ. The connectivity of networks, for example, changes drastically when λ passes a threshold value:

$$\lambda_{cr}(\kappa) = 1 - \sqrt[\kappa-1]{\frac{1}{\kappa}}. \tag{4}$$

The quantity κ in this equation represents the size of the alphabet. We have $\kappa = 4$ (**A,U,G,C**) for bases in single-stranded regions of RNA molecules and $\kappa = 6$ (**AU,UA,UG,GU,GC,CG**) for base pairs. Depending on the particular structure, the fraction of neutral neighbors is commonly different in the two subspaces of unpaired and paired bases and we are dealing with two parameter values, λ_u and λ_p, respectively. Neutral networks consist of a single component that spans the whole sequence space if $\lambda > \lambda_{cr}$, and below threshold $\lambda < \lambda_{cr}$ the network is partitioned into a large number of components, in general, a giant component and many small ones.

Exhaustive folding permits verification of the predictions of random graph theory and reveals further details of neutral networks. The typical series of components for neutral networks (either a connected network spanning whole sequence space or a very large component accompanied by several small ones) is indeed found with many common structures. There are, however, also numerous networks whose series of components are significantly different. We find networks with two as well as four equally sized large components, and three components with an approximate size ratio of 1:2:1. Differences between the predictions of random graph theory and the results of exhaustive folding have been readily explained in terms of special properties of RNA secondary structures,[12] (Figure 1).

Random graph theory, in essence, predicts that sequences forming the same structure should be randomly distributed in sequence space. Deviations from such an ideal neutral network can be identified as structural features that are not accounted for by nonspecific base pairing logic. All structures that cannot form additional base pairs when sequence requirements are fulfilled behave perfectly normally (class I structures in Figure 1). There are, however, structures that can form additional base pairs, and will generally do so under the minimum free-energy criterion,

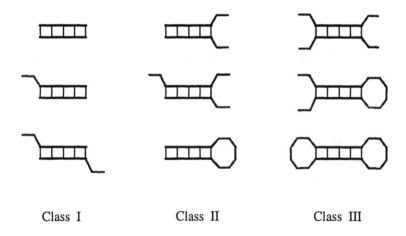

Class I   Class II   Class III

FIGURE 1   Three classes of RNA secondary structures forming different types of neutral networks. Class I structures contain no mobile elements (free ends, large loops or joints) or have only mobile elements that cannot form additional base pairs. The mobile elements of class II structures allow the extension of stacks by additional base pairs at one position. Stacks in class III structures can be extended in two positions. In principle, there are also structures that allow extensions of stacks in more than two ways, but they play no role for a short chain length ($n \leq 30$).

whenever the sequences carry complementary bases at the corresponding positions. Class II structures (Figure 1), for example, are least likely to be formed when the overall base composition is 50% **G** and 50% **C**, because here the probability for forming an additional base pair and folding into another structure is largest. If there is an excess of **G** ($\{50+\delta\}$%) it is much more likely that such a structure will actually be formed. The same is true for an excess of **C**, and this is precisely reflected by the neutral networks of class II structures with two (major) components: the maximum probabilities for forming class II structures are **G:C**=$(50+\delta)$:$(50-\delta)$ for one component and **G:C**=$(50+\delta)$:$(50-\delta)$ for the second one. By the same token, class III structures have two (independent) possibilities to form an additional base pair and, thus, they have the highest probability to be formed if the sequences have excess δ and ε. If no additional information is available, we can assume $\varepsilon \approx \delta$. Independent superposition then yields four equally sized components with **G:C** compositions of $(50+2\delta)$:$(50-2\delta)$, $2\times(50{:}50)$, and $(50-2\delta)$:$(50+2\delta)$, precisely as it is observed with four-component neutral networks. Three-component networks are *de facto* four-component networks in which the two central (50:50) components have merged to a single component. Thus, neutral networks are described well by the random graph model: the assumption that sequences folding into the same structure are randomly distributed in the space of compatible sequences is justified unless special structural features lead to systematic biases.

3. THE SEQUENCE-STRUCTURE MAP OF PROTEINS

The number of possible protein sequences is enormous. For $n = 100$ residues there are 20^{100} sequences. On the other hand, the repertoire of stable native folds seems to be highly restricted. As with RNA molecules, it has been frequently observed that seemingly unrelated protein sequences have essentially the same fold.[25] It seems, therefore, that RNA and proteins, despite their different chemistry, share fundamental properties of their sequence-structure maps.

Computational studies similar to the explorations of the RNA world reported above are precluded by the notorious complexity of the protein-folding problem, and by the fact that there is no biophysically meaningful and computationally simple coarse resolution of protein structures. (The term secondary structure refers, in the protein world, to local features that may or may not be present but do not capture the global organization of the molecule.) Hence, we have to resort to a less ambitious approach[1] based exclusively on inverse folding.

3.1 INVERSE FOLDING USING KNOWLEDGE BASED POTENTIALS

In order to characterize the topology of neutral sets of proteins $S(\psi)$, we need a technique for deciding whether a given sequence x is a member of $S(\psi)$, that is, whether x folds into the structure ψ. This problem is less demanding than predicting the unknown protein structure of a given amino acid sequence. It can be investigated by inverse folding techniques, (see e.g., Bowie et al.[2] and the references therein). In contrast to the RNA case, however, we cannot derive inverse folding from a solution to the protein-folding problem, as the latter is still unsolved.

The starting point is a potential function, $W(x, \psi)$, for evaluating the energy of a sequence x when folded into a structure ψ that is defined by the spatial coordinates of its C^α and C^β atoms. Recent studies using knowledge-based potentials demonstrated that the energy of the native fold (i.e., putative ground state) of a sequence x can be estimated from the distribution of the energy values of x in its conformation space. This allows the construction of an energy scale by which conformations of different sequences can be compared. As a measure for the quality of fit of sequence x and structure ψ, the *z-score*[3]

$$z(x, \psi) = \frac{W(x, \psi) - \overline{W}(x)}{\sigma_W(x)} \tag{5}$$

is used. Here $\overline{W}(x)$ is the average energy of sequence x in all conformations in a database, and $\sigma_W(x)$ is the standard deviation of the corresponding distribution. Empirically, native folds have z-scores in a narrow characteristic range. Furthermore, use of the *z-score* introduces a kind of negative design that avoids sequences with multiple stable structures. The z-score correlates well with the rms deviation of alternative structures from the native structure. The computed z-score also improves with increasing resolution of the X-ray structure, when computed for the same protein.[1] The **PROSA II** potentials[3,33] are particularly suitable for studying neutral sets in protein space.[2]

Hence, one may assume that x is a member of $S(\psi)$ if the z-score of x in conformation ψ is in the native range.[3] Of course, only native structures ψ that are already in the database can be explored by this method. Formally, we have translated inverse folding into an optimization problem on the set of all sequences: we are looking for the minima x of the z-score $z(x, \psi)$. From the computational point of view, this optimization problem appears to be very easy. Indeed, it is sufficient to use the simplest heuristic, the *adaptive walk*, which repeatedly tries random mutations (exchanges of single amino acid) that are accepted if and only if the z-score decreases.

Whether the sequences predicted by this inverse-folding procedure do indeed fold to the desired structure can ultimately only be answered by experiment. Independent criteria, however, would at least indicate whether the assumption is

---

[2] For the space of amino acid sequences we use the synonym *protein space*.[23]

reasonable. One finds, for instance, that the SOPM and PHD predictions of secondary structures of inverse-folded sequences agree with the target secondary structure[1] despite the fact that the PROSA II potentials make no explicit reference to secondary structure.

3.2 DISTRIBUTION OF INVERSE-FOLDED PROTEINS

Amino acid sequences generated by independent adaptive walks show little or no similarity to the wild-type sequence or among each other. This is consistent with the observation that a significant sequence homology is not necessary for two proteins to have a common fold. Although they lie somewhat closer together than random sequences with a typical amino acid composition (taken from the SwissProt database), pairs with the maximum Hamming distance do occur. Tree reconstruction methods, such as neighbor joining and the split-decomposition technique, suggest that the sequences with wild-type-like z-scores are distributed essentially randomly in sequence space.

A number of groups have argued that the pattern of hydrophobic versus hydrophilic amino acids (**HP**-pattern) has a dominating influence on protein structure.[6,18] Inverse-folded sequences are very flexible at the level of individual amino acids but require a significant level of conservation of amino acid classes. Hence, it is natural to ask whether all 20 amino acids are, in fact, necessary to build native protein structures, or whether this can be done with a (small) subset of different amino acids. Not surprisingly, no sequences with wild-type-like structures could be found when only hydrophilic amino acids or only hydrophobic amino acids were used. Surprisingly, however, we observed substantial differences between alphabets that all contain both hydrophilic and hydrophobic amino acids. For instance, the two-letter alphabet **AD** gives very poor results, while other combinations of just one hydrophilic and one hydrophobic amino acid, such as **LS** or **DL**, yield wild-type-like z-scores. It is not surprising that **ADL** and **ADLG** yield good sequences since **DL** is already sufficient. The inverse-folded sequences in these alphabets do, however, contain a substantial fraction of **A** and **G**. The alphabet **ADLG** has been proposed as a candidate for a primordial set of amino acids, even before the full genetic code has been developed. It is reassuring to see that this alphabet allows inverse folding of a variety of present-day protein structures. It is worth noting in this context that the **QLR** alphabet used in experimental work on random polypeptides by Sauer and coworkers[4] does not yield wild-type-like z-scores for globular protein structures. This may not be surprising since Sauer's experimental **QLR**-peptides form multimeric structures.

3.3 NEUTRAL NETWORKS IN PROTEIN SPACE

Inverse-folded sequences with z-scores below the threshold z^* were used as starting points for neutral paths. The substitution frequencies for the production of mutants were computed from the natural frequencies of the amino acids as contained in the SwissProt database.

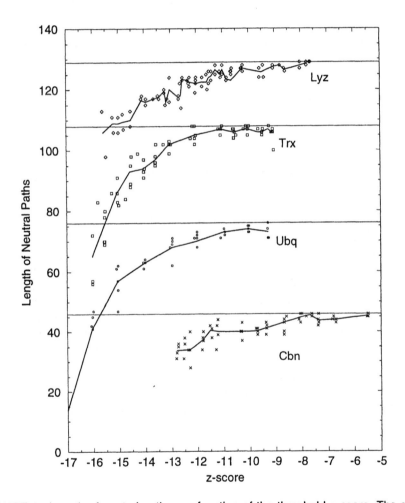

FIGURE 2 Length of neutral path as a function of the threshold z-score. The solid lines indicate the averages over the available data for each protein. The chain lengths of the four proteins 1cbn, 1ubq, 2trxA, and 1lyz are indicated by thin solid lines for comparison. The rightmost data points for each structure correspond to the wild-type z-scores.

TABLE 3 Characteristics of protein neutral networks.

Protein	n	z	$\mathcal{L}^1$	L_{adap}	$\langle d \rangle_{adw}$	$\langle d \rangle_{nn}$
1cbn	46	−5.50	44.6	17.7	38.7	42.3
1ubq	76	−9.30	72.5	61.9	61.1	66.3
2trxA	108	−9.22	106.3	71.7	87.7	97.5
1lyz	129	−7.70	126.2	58.2	106.0	118.7

[1] The length of neutral path is averaged over all data with z-scores between wild type and three standard deviations better than wild type.

Figure 2 shows the results for four different protein structures.[1] The lengths of the neutral paths $\mathcal{L}$ are roughly equal to the lengths n of the proteins at z-score levels comparable to the wild-type sequence. Even at z-scores about six standard deviations better than the wild-type z-score, the length of the neutral paths is still greater than three quarters of the length of the protein. The average values of $\mathcal{L}$ taken over the z-score interval $z_{w.t.} - 3 \leq z \leq z_{w.t.}$ are collected in Table 3. The average Hamming distances between the endpoints of unrelated neutral paths, $\langle d \rangle_{nn}$, are in the range of 90 to 95% of the chain length, indicating that the neutral networks span essentially the entire sequence space. It is not surprising that the Hamming distances between the end points of neutral paths is somewhat larger than the average distance, $\langle d \rangle_{adw}$, between the end points of adaptive walks, since a neutral walk has a built-in bias toward sequences with a more uniform distribution of amino acids.

Extensive studies on neutral paths in restricted protein alphabets have been reported only for the "primordial alphabet" **ADLG**.[1] The average length of neutral paths in this alphabet is 76% to 87% of the sequence length. Note that the distance between random sequences is 75% of the chain length in a four-letter alphabet. The neutral paths thus extend well beyond the mean distance of random sequences even in some highly restricted amino acid alphabets.

4. ADAPTATION DYNAMICS ON REDUNDANT LANDSCAPES

Landscapes assign fitness values to genotypes, i.e., to individual polynucleotide sequences in sequence space. Based on (point) mutation (and recombination) being

Adaptive Walks without Selective Neutrality

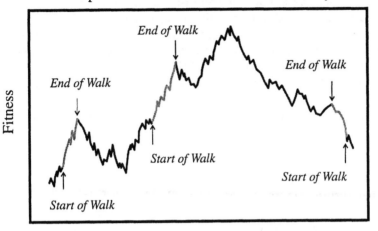

Sequence Space

Adaptive Walk on Neutral Networks

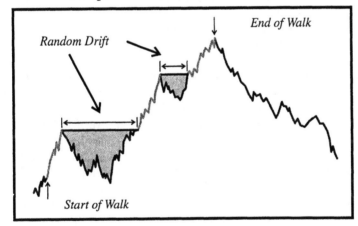

Sequence Space

FIGURE 3 The role of neutral networks in evolution. Optimization occurs through adaptive walks and random drift. Adaptive walks provide a choice for the next step arbitrarily from all directions where fitness is (locally) nondecreasing. Populations can bridge narrow valleys with widths of a few point mutations. In the absence of selective neutrality (upper part) they are, however, unable to span larger Hamming distances and thus will approach only the next major fitness peak. Populations on rugged landscapes with extended neutral networks evolve along the networks by a combination of adaptive walks and random drift at constant fitness (lower part). In this manner, populations bridge over large valleys and may eventually reach the global maximum of the fitness landscape.

the evolutionarily adequate move set and distance measure, fitness landscapes are highly rugged in the sense that they contain a high number of local optima on all scales, (see Stadler[34] and the references therein). Populations optimize mean fitness by migration through sequence space. In the absence of neutrality, populations climb on landscapes until they reach one of the minor peaks where migration ends because all surrounding genotypes have lower fitness (Figure 3). Neutral networks of RNA molecules play an important role in evolutionary optimization, as they enable populations to escape from evolutionary traps in the form of local fitness optima.

On neutral networks and, likewise, on flat landscapes, populations migrate by a diffusionlike mechanism.[5,16] Whenever adaptive migration ends on a local fitness optimum, the populations start to drift on the neutral network belonging to the structure that corresponds to this optimum. Random drift is continued until the population reaches an area in sequence space where some fitness values are higher than that of the network. Then another period of adaptive evolution sets in. A complete optimization run thus appears as a stepwise process: phases of increasing mean fitness are interrupted by "static" periods with mean fitness values fluctuating around a constant value (Figure 4). When the network belongs to a common structure its extension through whole sequence space may eventually allow the population to find the global optimum.

Optimization of protein structures requires translation of genes into protein. Phage display or bacterial display, for example, provides an excellent tool for *in vitro* evolution of proteins. The existence of extensive neutral networks meets a claim raised by Maynard-Smith[23] for protein spaces that are suitable for efficient evolution. Empirical evidence for a large degree of *functional* neutrality in protein space has been presented recently by Wain-Hobson and coworkers.[22]

The scenario described above has been studied and verified by a series of computer simulations.[9,16] A very detailed study on RNA optimization has been performed by means of a simple model of molecular evolution in which the fitness of a sequence expressed in terms of its replication rate depends on the structure. The evolution of a population of RNA sequences of fixed chain length n is simulated under conditions of a flow reactor that adjusts the total population size to fluctuate around a constant capacity N. Individual sequences replicate with a structure-dependent rate constant that is defined to be a function of the distance between its secondary structure and a predefined target structure. Mutation is introduced by simulating a copying mechanism that copies each base with fidelity $1 - p$. In this simulation, as well as in evolution *in vitro* and *in vivo*, there are two sources of neutrality: one is the sequence-to-structure mapping discussed here, the other is the structure-to-replication rate (fitness) mapping.

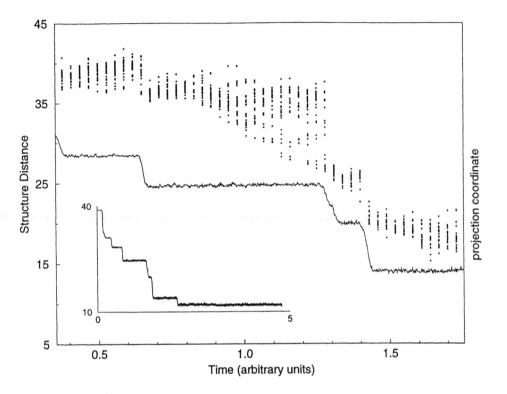

FIGURE 4 The stepwise course of evolutionary optimization of RNA structures.[16] A flow reactor with capacity $N = 1000$ is initialized with that many copies of a random sequence of length $n = 76$. The mutation rate is $p = 0.001$ and the target secondary structure is the tRNA[Phe] clover leaf; the replication rate function is $A(D) = 1.06^{146-D}$ where D is the tree-edit distance to the target structure. The population average of the distance to the target is plotted against time (full line) for a specific interval of the entire run (shown in the inset). The scattered points in the upper part of the plot indicate the position of the population in sequence space as a projection to a single coordinate. The fitness plateaus correspond to diffusion on neutral networks. The population spreads out in sequence space during these periods. Sudden jumps indicate the transition to another network. Darwinian selection drastically reduces the sequence diversity at the transition points.

Approximating the dynamics of the flow reactor using the Moran model, in which the (error-prone) replication of a randomly chosen sequence is followed by the removal of a randomly chosen sequence, the diffusion coefficient D_0 can be computed for the flat landscape. One finds

$$D_0 = \frac{6anp(1 + 1/N)}{3 + 4Np} \approx \frac{6anp}{3 + 4Np},$$

where a is the common replication rate.[16] For small mutation rates the diffusion coefficient, D, on the structure-dependent landscape can be approximated by $D = D_0 \bar{\lambda}$, where $\bar{\lambda}$ denotes the average fraction of neutral mutants for the dominant structure. The diffusion of finite populations in sequence space is directly related to Kimura's neutral theory,[20] which stresses a different aspect, namely the number of nucleotide substitutions that reach fixation per generation k, also referred to as the "rate of evolution." The theory yields $k = a\,p\,\nu\,\bar{\lambda}$, and hence $D = 6\,k/(3 + 4Np)$, for small mutation rates.

Studies of evolutionary dynamics on rugged model landscapes which did not involve a genotype-phenotype model showed a loss of sequence information at a critical error rate corresponding to the loss of the dominant phenotype.[8] In the presence of percolating neutrality, however, all sequence information is lost at any nonzero error rate due to diffusion and the finiteness of the population. Yet there is another threshold, the *phenotypic error threshold* p_c, beyond which the dominant phenotype is lost as well. That is when evolutionary adaptation breaks down.

Diffusion in sequence space, the connection with Kimura's neutral theory, and the phenotypic error threshold are consequences of the existence of neutral networks. Shape-space covering implies a constant rate of innovation[17]: while diffusing along a neutral network, a population constantly produces nonneutral mutants folding into different structures. Shape-space covering implies that almost all structures can be found somewhere near the current neutral network. Hence, the population keeps discovering structures that it has never encountered before at a constant rate. When a superior structure is produced, Darwinian selection becomes the dominating effect and the population "jumps" onto the neutral network of the novel structure while the old network is abandoned (Figure 4).

5. DISCUSSION

Genotype-phenotype mappings of RNA molecules have been investigated using a prediction algorithm for secondary structures and by means of inverse folding. RNA secondary structures and protein folds are highly redundant in the sense that many sequences fold into the same shape. The most striking similarity in sequence-structure mappings of RNA molecules and proteins is the existence of few common and many rare structures. RNA secondary structures from **GC** sequences of chain length $n = 27$ yield a log-rank/log-frequency of occurrence-plot that almost coincides with the corresponding plot derived from **HP**-lattice proteins[3] (for comparison see Grüner et al.,[11] and Li,[21] and Figure 5). Such a distribution of

---

[3]By this expression we characterize protein models that distinguish only hydrophilic and hydrophobic residues. Sequences are folded on lattices to yield stable folds according to an energy criterion.[6] In the present case, the energy equals the number of **HH** contacts. The configurations are restricted to self-avoiding walks in a $3 \times 3 \times 3$ cube, hence $n = 27$.

the frequencies of structures can be characterized by a generalization of Zipf's law. In the example discussed here, **GC** sequences of chain length $n = 30$, more than 93% of all sequences fold into only 10% of all structures. Extrapolation of our data to longer chains indicates an increasing percentage of sequences folding into a decreasing fraction of structures. There are also strong indications that less coarse notions of structure give rise to very similar frequency distributions. Implications for evolutionary optimization are evident: populations, in essence, live in a space of common structures or phenotypes. Rare phenotypes are extremely hard to find in random searches and thus play no role in evolution. Still, one important feature for understanding evolution is missing: we do not know the rules that determine whether a phenotype is common or rare. In other words, given the structure of a phenotype, we should be able to predict the fraction of genotypes that fold into it.

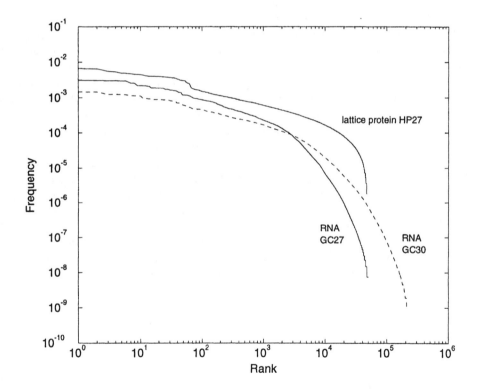

FIGURE 5   The distribution of frequencies of RNA secondary structures and lattice models of proteins. The diagram shows the distribution of preimage sizes of sequence-structure mappings for **GC** sequences of length 27 and 30[11] compared with the analogous plot computed for **HP** lattice proteins of chain length 27.[21]

In this respect, investigations aiming at such a completion of the current concept, are under way. First results show that the modular building principle of natural biopolymers is highly important for the probability of realization of structures.

Algorithms for folding RNA sequences into secondary structures as well as inverse-folding procedures for proteins based on knowledge-based potentials predict extended connected networks of sequences with identical structure. In addition, RNA secondary structures exhibit shape-space covering, that is, any common structure can be found within a small radius in sequence space. (It not yet clear whether protein space shares this property.) These observations have striking consequences for adaptation, based on a fairly realistic model of test-tube evolution: (1) Finite populations diffuse along neutral networks where their dynamics confirm the predictions of Kimura's neutral theory. After a sufficiently long period of time (set by the diffusion coefficient) all sequence information is lost, yet the phenotype is conserved. It is the maintenance of a phenotype, not of a genotype, which defines the mutation threshold beyond which adaptation breaks down. (2) On a single neutral network the population splits into well-separated clusters. A population is not a single localized quasispecies in sequence space,[8] but rather a collection of different quasispecies. Each undergoes independent diffusion, while all share the same dominant phenotype. (3) Neutral networks of different structures are interwoven. While drifting on a neutral network, a population produces a fraction of mutants off the network and thereby explores new phenotypes. A selection-induced transition between two structures occurs in regions of sequence space where their networks come close to one another. The independent diffusion of subpopulations increases the likelihood that a population encounters such transition regions.

Neutral evolution, therefore, is not a dispensable addendum to evolutionary theory as it has often been suggested. On the contrary, neutral networks, arising as a consequence of the redundancy of sequence-structure (and possibly also structure-function) relationships of biopolymers, provide a powerful mechanism through which evolution can become truly efficient.

This is of particular importance for RNA virus evolution where, in favorable cases like influenza A or HIV, data are available not only for conventional phylogenetic trees but also on sequence variation within a single infected individual. Many conserved secondary structure elements, such as the TAR hairpin or the five-fingered motif of the RRE region in HIV, have been identified as important regulatory elements—yet their underlying sequences exhibit a substantial number of (compensatory) mutations.

ACKNOWLEDGMENTS

Extensive discussions on the subject of this review with Drs. Walter Fontana, Christian Forst, and Martijn Huynen are gratefully acknowledged. The work on inverse-protein folding is joint research with Mag. Aderonke Babajide, Dr. Ivo L. Hofacker, and Prof. Manfred Sippl. The work was supported financially by the Austrian *Fonds zur Förderung der Wissenschaftlichen Forschung*, Projects No. 10578-MAT and 11065-CHE; by the European Commission, Contract Study PSS*0884; by the Diversity Biotechnology Consortium (New Mexico); and by the Santa Fe Institute.

REFERENCES

1. Aderonke, B., and I. L. Hofacker, M. J. Sippl, and P. F. Stadler. "Neutral Networks in Protein Space: A Computational Study Based on Knowledge-Based Potentials of Mean Force." *Folding & Design* (1997): In press. Also published as Santa Fe Institute Working Paper No. 96-12-085.
2. Bowie, J. U., R. Luthy, and D. Eisenberg. "A Method to Identify Protein Sequences that Fold into a Known Three-Dimensional Structure." *Science* **253** (1991): 164–170.
3. Casari, G., and M. J. Sippl. "Structure-Derived Hydrophobic Potentials—Hydrophobic Potentials Derived from X-Ray Structures of Globular Proteins Is Able to Identify Native Folds." *J. Mol. Biol.* **224** (1992): 725–732.
4. Davidson, A. R., and R. T. Sauer. "Folded Proteins Occur Frequently in Libraries of Random Amino Acid Sequences." *Proc. Natl. Acad. Sci., USA* **91** (1994): 2146–2150.
5. Derrida, B., and L. Peliti. "Evolution in a Flat Fitness Landscape." *Bull. Math. Biol.* **53** (1991): 355–382.
6. Dill, K. A., S. Bromberg, K. Yue, K. M. Fiebig, D. P. Yeo, P. D. Thomas, and H. S. Chan. "Principles of Protein Folding: A Perspective from Simple Exact Models." *Prot. Sci.* **4** (1995): 561–602.
7. Eigen, M., R. Winkler-Oswatitsch, and A. W. M. Dress. "Statistical Geometry in Sequence Space: A Method of Comparative Sequence Analysis." *Proc. Natl. Acad. Sci., USA* **85** (1988): 5913–5917.
8. Eigen, M., J. McCaskill, and P. Schuster. "The Molecular Quasispecies." *Adv. Chem. Phys.* **75** (1989): 149–263.
9. Fontana, W., and P. Schuster. "A Computer Model of Evolutionary Optimization." *Biophys. Chem.* **26** (1987): 123–147.
10. Fontana, W., D. A. M. Konings, P. F. Stadler, and P. Schuster. "Statistics of RNA Secondary Structures." *Biopolymers* **33** (1993): 1389–1404.
11. Grüner, W., R. Giegerich, D. Strothmann, C. Reidys, J. Weber, I. L. Hofacker, P. F. Stadler, and P. Schuster. "Analysis of RNA Sequence Structure Maps by Exhaustive Enumeration, I: Neutral Networks." *Monath. Chem.* **127** (1996): 355–374.
12. Grüner, W., and R. Giegerich, D. Strothmann, C. Reidys, J. Weber, I. L. Hofacker, P. F. Stadler, and P. Schuster. "Analysis of RNA Sequence Structure Maps by Exhaustive Enumeration, II: Structures of Neutral Networks and Shape-Space Covering." *Monath. Chem.* **127** (1996): 375–389.
13. Haslinger, C., and P. F. Stadler. "RNA Secondary Structures with Pseudoknots." (1997).
14. Hofacker, I. L., W. Fontana, P. F. Stadler, S. Bonhoeffer, M. Tacker, and P. Schuster. "Fast Folding and Comparison of RNA Secondary Structures." *Monatsh. Chem.* **125** (1994): 167–188.

15. Hofacker, I. L., M. A. Huynen, P. F. Stadler, and P. E. Stolorz. "Knowledge Discovery in RNA Sequence Families of HIV Using Scalable Computers." In *Proceedings of the 2nd International Conference on Knowledge Discovery and Data Mining*, edited by E. Simoudis, J. Han, and U. Fayyad, 20–25. Portland, OR: AAAI Press, 1996.

16. Huynen, M. A., P. F. Stadler, and W. Fontana. "Smoothness within Ruggedness: The Role of Neutrality in Adaptation." *Proc. Natl. Acad. Sci., USA* **93** (1996): 397–401.

17. Huynen, M. A. "Exploring Phenotype Space Through Neutral Evolution." *J. Mol. Evol.* **43** (1996): 165–169.

18. Kamtekar, S., J. M. Schiffer, H. Xiong, J. M. Babik, and M. H. Hecht. "Protein Design by Binary Patterning of Polar and Nonpolar Amino Acids." *Science* **262** (1993): 1680–1685.

19. Kauffman, S. K. *The Origins of Order: Self-Organization and Selection in Evolution.* Oxford, UK: Oxford University Press, 1993.

20. Kimura, M. *The Neutral Theory of Molecular Evolution.* Cambridge, UK: Cambridge University Press, 1993.

21. Li, H., R. Helling, C. Tang, and N. Wingreen. "Emergence of Preferred Structures in a Simple Model of Protein Folding." *Science* **273** (1996): 666–669.

22. Martinez, M. A., V. Pezo, P. Marliére, and S. Wain-Hobson. "Exploring the Functional Robustness of an Enzyme by *In Vitro* Evolution." *EMBO J.* **15** (1996): 1203–1210.

23. Maynard-Smith, J. "Natural Selection and the Concept of a Protein Space." *Nature* **225** (1970): 563–564.

24. McCaskill, J. S. "The Equilibrium Partition Function and Base Pair Binding Probabilities for RNA Secondary Structure." *Biopolymers* **29** (1990): 1105–1119.

25. Murzin, A. G. "Structural Classification of Proteins: New Superfamilies." *Curr. Opin. Struct. Biol.* **6** (1996): 386–394.

26. Ohta, T. "Population Size and Rate of Evolution." *J. Mol. Evol.* **1** (1972): 305–314.

27. Rauscher, S., C. Flamm, C. Mandl, F. X. Heinz, and P. F. Stadler. "Secondary Structure of the 3'-Non-Coding Region of Flavivirus Genomes: Comparative Analysis of Base Pairing Probabilities." *RNA* **3** (1997): 779–791. Also Santa Fe Institute Working Paper 97-02-010.

28. Reidys, C., P. F. Stadler, and P. Schuster. "Generic Properties of Combinatory Maps—Neutral Networks of RNA Secondary Structures." *Bull. Math. Biol.* **59** (1997): 339–397.

29. Schuster, P., W. Fontana, P. F. Stadler, and I. L. Hofacker. "From Sequences to Shapes and Back: A Case Study in RNA Secondary Structures." *Proc. Roy. Soc. (London) B* **255** (1994): 279–284.

30. Schuster, P., and P. F. Stadler. "Landscapes: Complex Optimization Problems and Biopolymer Structures." *Computers & Chem.* **18** (1994): 295–314.

31. Schuster, P. "How to Search for RNA Structures: Theoretical Concepts in Evolutionary Biotechnology." *J. Biotechnology* **41** (1995): 239–257.
32. Schuster, P., P. F. Stadler, and A. Renner. "RNA Structure and Folding: From Conventional to New Issues in Structure Predictions." *Curr. Opin. Struct. Biol.* **7** (1997): 229–235.
33. Sippl, M. J. "Calculation of Conformational Ensembles from Potentials of Mean Force—An Approach to the Knowledge-Based Prediction of Local Structures in Globular Proteins." *J. Mol. Biol.* **213** (1990): 859–883.
34. Stadler, P. F. "Towards a Theory of Landscapes." In *Complex Systems and Binary Networks*, edited by R. Lopéz-Peña, R. Capovilla, R. García-Pelayo, H. Waelbroeck, and F. Zertuche, 77–163. Berlin; New York: Springer-Verlag, 1995.
35. Tacker, M., P. F. Stadler, E. G. Bornberg-Bauer, I. L. Hofacker, P. Schuster. "Algorithm Independent Properties of RNA Secondary Structure Predictions." *Eur. Biophys. J.* **25** (1996): 115–130.
36. Waterman, M. S. "Secondary Structure of Single-Stranded Nucleic Acids." *Adv. Math. (Suppl. Studies)* **1** (1978): 167–212.
37. Zuker, M., and D. Sankoff. "RNA Secondary Structures and Their Prediction." *Bull. Math. Biol.* **46** (1984): 591–621.

Dr. Simon Wain-Hobson
URM, Pasteur Institute, 25 Rue du Dr. Roux, 75724 Paris Cedex 15, FRANCE;
E-mail: simon@pasteur.fr

Plurality in HIV Genetics

It seems that a contemporary cultural defect has been allowed to color our thinking—monomania. Let's take a recent example which few could have missed—the centenary Olympic Games in Atlanta. For example, Nike's unambiguous advertisement—you don't win silver, you lose gold!? Given an advance of say 1/100th of a second, do we know mechanistically what that means? Sometimes we are in the realm of stochastic factors such as a gust of wind and, logically, its orientation. Of course, there were more obvious wins such as the fabulous men's 200 meters. Yet how many of us can name one athlete who came fourth in a final? While we can't devote brain cells to every piece of information, the truth is that that individual was the fourth best in the world at that time and place. Fourth among how many billions? Surely that's something? Another oddity is the obvious hierarchy in gold medals—clay pigeon shooting just doesn't rival track events, leading to the inescapable conclusion that some gold medals are more equal than others. Yet didn't Baron Pierre de Coubertin revive the Games to foster the Olympic spirit—"to have participated"?

The number of Olympic gold medalists who didn't perform well in the Zurich athletic meeting the following month was surprising. Sergei Bubka, unquestionably the greatest pole vaulter by the huge margin of 10–15 cm, just didn't go anywhere in Atlanta. Yet in Zurich he came out on top. It is as though excellence exists as a surface rather than a point on a rugged landscape where the average altitude distinguishes the really excellent from the so-so. A partial analogy may be seen in the difference between climate and weather—for a given region climate is the temporal average of weather. Weather can be very variable, climate much less so. And as we all know there are better moments in our scientific careers.

And so it seems we have allowed too simplistic notions to intrude into our understanding of microbes—RNA viruses and retroviruses for the purposes of this diatribe. The mutation rates of RNA viruses approach something like 0.1–2 mutations per genome per cycle, which are in sharp contrast to those for DNA-based microbes like HPV, herpes viruses, and bacteria.[6,7,9,10,12,17,19] This difference is due to the fact that RNA and retroviral replication complexes are devoid of a 3′ exonucleolytic activity. Consequently any RNA viral sample is composed of a collection of related, but subtly distinct, genomes. Such a collection of viral genomes is called a viral quasispecies.[13] A veritable deluge of sequence data has amply highlighted such complexity.

An elegant theoretical description of RNA viruses has been provided by Manfred Eigen and colleagues.[13] With clarity they describe the properties of such ensembles of genomes and use the term quasispecies to capture their quixotic nature. It is essentially a deterministic model. Note that the description pertains to an infinite array of freely diffusing genomes, there being no spatial discontinuities or barriers. In addition there is no viral predator, such as the immune system; the genomes are merely competing for resources. Any virologist will recognize that such a description approximates to a viral culture. Beautiful results ensuing from the long-standing collaboration between John Holland and Esteban Domingo have, more than others, borne this out.[7,11,19,25,26,31]

Monoclonal antibody escape mutants and drug-resistant forms are readily selected. RNA viruses are so sensitive to the surrounding ecological niche that they can distinguish between different clones of the same parent cell line.[5] Exponential fitness gains have been noted upon large populations passages of vesicular stomatitis virus (VSV) strains.[25] In competition studies using pairs of marked VSV populations, fitness gains have been noted for both throughout the multiple-passage history. These data support the notion that the resulting genomes "are the best of the best of the best."[18]

And so they are for tissue culture niches. Yet to what extent does such a degree of competition exist in the real world? The crucial differences between tissue culture and the natural host are (1) the existence of spatial heterogeneity; (2) the presence of an immensely powerful predator, alias the immune system; and (3) the population of genomes *in vivo* is vastly smaller than the sequence space available to a virus.

SPACE. A glance at any histology slide or textbook is a salient reminder of spatial discontinuities over distances of one or two cell diameters. For HIV and the hugely delocalized immune system with a mêlée of different lymphoid organs, a myriad of subtly different susceptible cell types, diffusion of viral progeny is certainly restricted. The exquisite spatial heterogeneity of HIV within splenic white pulps has been described.[3] For HPV infiltration of skin, spatial discontinuities and gradients are apparent. Furthermore, spatial differences allow allopatric speciation. On a different note, is it reasonable to assume that the British could have resisted Napoleon or Hitler without the English Channel?

PREDATORS. The power of the immune system should never be underestimated, vaccination probably being the most poignant reminder. Three examples from HIV/AIDS further emphasize this point. (1) The outgrowth of opportunistic infections is inversely related to the CD4 T cell count. These cells are crucial to coordination of adequate immune responses. (2) Based on the rate of SIV sequence divergence from a known origin, it has been concluded that the vast majority of progeny ($> 99.9\%$) are cleared, being unable to give rise to a productively infected cell.[28] (3) The dynamics of HIV clearance have recently been highlighted following "perturbation" of virus replication by powerful triple therapy. It has been estimated that between 10^7–10^{10} virions may be cleared daily from the peripheral blood.[16,29,34]

FINITE POPULATION SIZES. A number of points can be made.

1. VSV, HIV, HCV, and FMDV field isolates all exhibit considerable genetic variation. Presently, HIV-1 envelope protein variation is approaching 30–40%. Intrapatient variation for the same region can approach 10% within ten years or so. Viruses from no two patients are strictly alike. These values dwarf the variation among primate species, which are evolving some 10^5–10^6 times more slowly.[15] Between strains, HPVs are highly divergent, although, due to host cell proofreading, their mutation rate is much lower than that of HIV and other RNA viruses. Nevertheless, as John Maynard Smith noted for genomes with genome mutation rates <1, all the single mutation intermediates separating two observable sequences must themselves be viable.[23] Thus, even for the slow moving papillomaviruses the sequence space available to them is huge. Therefore, any observable quasispecies represents a trivial fraction of viral sequence space.
2. For proteins, third base codon degeneracy and conservative amino acid substitutions suggest a degree of redundancy, although it would be clearly wrong to suggest that all such substitutions are neutral. Be that as it may, saturation mutagenesis and *in vitro* evolution of proteins reveals a large number of viable protein variants, sometimes with improved kinetic constants.[27]
3. The starting position of a genome in sequence space is important because mutation and back mutation frequencies are not necessarily equal (Figure 1). The

G→A forward transition results from $G_{template}$:dTTP mispairing while the back mutation could result from either a $C_{template}$:dATP or $T_{template}$:dGTP. Conan Doyle furnishes us with an analogy. A criminal from Chicago tried to warn Sherlock Holmes of the consequences of meddling in his affairs by bending a steel poker with his bare hands. Holmes rose to the occasion by straightening out the bent poker—the more difficult feat. Furthermore, it is well known that the two base pairs preceding the mismatch site strongly influence the Km and Vmax of the substitution, sometimes by up to two orders of magnitude. Suppose that, in the initial experiment, a fitness gain followed from a G:T mismatch in the context GpA (+ strand as reference) leading to ApA (Figure 1). Suppose that, in the rerun with a different reference, the mutation was in the context of GpT. The probability of the mutation arising could be hugely different, allowing selection of a variant with a mutation elsewhere in its genome.

4. A number of different proteins and nucleic acid elements can contribute to overall replication fitness. For example, deletions in virtually all of the HIV/SIV genes reduces or abolishes replication.[22] Is there a single optimum for the ensemble or a series of compromises resulting from the high dimensionality of having 19 mature proteins and at least 6 RNA secondary structures? We don't know, yet the above example suggests caution before opting for the single solution.

5. Schuster and colleagues have shown that the constraints of a stable nucleic acid stem-loop structure still permit a truly vast number of combinations. Furthermore, they show that it is possible to pass readily from one to another, i.e., that such structures do not exist as discrete and isolated solutions.[20,32]

6. Szostak's group has shown that, through repeated rounds of selection, it is possible to pull out a ribozyme from a sample of randomly synthesized RNA molecules.[1,14] Apart from being a tour de force and probably the most flagrant disproof of any "force vitale," they estimated that the frequency of such events was of the order of 10^{-14}. The reciprocal of such a frequency is trivial given the enormity of sequence space. There are clearly many ways to be a ribozyme.

In deterministic-based models such as the viral quasispecies, fitness is omnipresent. Such models require an infinite (read very large) population of virions to freely diffuse and therefore compete. *In vivo* only natural selection over time can sift through the immensity of viral sequence space. However, spatial discontinuities and the bottlenecking resulting from massive destruction work against fitness selection because they reduce the chance for competition between variants. When populations are small and progeny are not competing efficiently, it is not obvious how a virus will evolve. Local founder effects take on particular importance. All this emphasizes that there are many ways of being a virus, that viral sequence space is large and extensively occupied in the real world. There is abundant evidence for plurality in the RNA viral world. One problem, however, is that plurality doesn't fit the one-dimensional mold of the molecular biologist nor our global economy, which clamors for winners. It is complex, more difficult to manage.

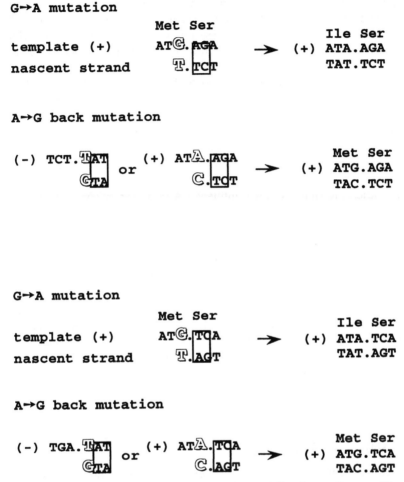

FIGURE 1   Back and forth. The frequency of the forward G→A mutation resulting from a $G_{template}$:T mismatch is strongly dependent on the nature of the two base pairs preceding it (boxed). The A→G back mutation may result either from an $A_{template}$:C or $T_{template}$:G mismatch. The same mutations in a different context (lower part), yet one that is synonymous in terms of the genetic code, involve different boxed dinucleotides which translate into different mutation probabilities.

Negative selection, the winnowing out of the obviously defective or poorly replicating genomes (at least with respect to the input virus), is clearly operative. This follows from the finding of an excess of synonymous substitutions either in cross-sectional studies of HIV isolates[24] or from longitudinal intrapatient studies of certain SIV/HIV genes such as *int* (Johnson et al.[21]) or *vif* (Sova et al.[33]). It also follows from the description of hypervariable and conserved regions within the HIV envelope sequences. Other longitudinal studies of the hypervariable regions in *env* and *nef* have shown genetic drift over a scale of a few months to years.[28,30] Presumably the selection pressures on these regions are markedly weaker than on *int* and *vif*, meaning that in the time span under study, it is sometimes hard to pick up the signal.

Many attempts have been made to find out whether HIV, or other viruses, escape specific immune responses by ongoing mutation. Be this as it may, it has been remarkably difficult to find good examples. When found, they are invariably associated with a usual focusing of the immune system on a single epitope.[2] Such cases are rare. Note that this is distinct to the influenza A paradigm for genetic variation. Here novel mutations allow variants to infect partially immunocompetent hosts. Despite a genomic mutation rate some 3–5 fold greater than HIV,[10] there is no data indicating that, de novo, influenza A variation is necessary to sustain the infection. The same is true of other RNA viral infections.

Paraphrasing the creator of Lucky Luke, why is it that RNA viruses, capable of mutating faster than their shadow, do not seem to exploit this resource? Does this mean that positive selection is not relevant to viral evolution? Surely the viral world can't be drifting? This question appears threatening because it would seem to strike at our Darwinian view of the world. It certainly does not, although sifting through the vast volumes HIV sequence data, it is clear that negative selection and plurality predominate. All it means is that viral sequence space is large, degenerate, and that a large proportion of observable mutations are not seen by natural selection. Nevertheless, there are a number of additional facets worth considering, notably those of time scales, frequencies, and perception.

TIME AND FREQUENCIES

By definition, acute viral infections are over within a few weeks. During this time the virus has an initial advantage since mounting antigen-specific immunity requires some time. As it kicks in, the titres of viremia and virus-producing cells decline to extremely low values. As the severity of clinical symptoms are invariably related to virus load, the individual recovers. Note that in this short time, the number of consecutive rounds of replication are small—for a six-hour cycle (e.g., as for VSV) three weeks corresponds to ~80 cycles. Only for variants with very large selection coefficients (relative replication with respect to input virus) would there be sufficient time to allow their outgrowth. Furthermore, in a population of variants, the fitter variant will generally be outnumbered. As they compete for resources, competitive exclusion will also occur, a phenomenon that has been shown for cultured viruses.[4]

The same is true for persistent viral infections. Their survival strategy has to be set up before the monstrously powerful viral-specific immunity has been mounted

and refined. (Monstrous? A relative term of course. Put it this way, one can re-
cover from acute hepatitis, synonymous with the eradication of billions of virions
and infected cells.) As mentioned above, there is rarely enough time to select for a
viral variant with anything but a huge fitness gain. So their strategy for persistence
must either involve interfering with the biology of the immune system or simply
subverting it by seeking out those "privileged" sites, such as the brain and kid-
neys, where the immune system is less powerful than in other organs and tissues.
Alternatively, given that natural populations are invariably outbred, there may be
subpopulations whose immunity is fractionally weaker than average. Then there are
the immunodeficient.

PERCEPTION

The need for a winner has been mentioned. (Incidentally this is in patent contra-
diction with our experience of promotions within government or private research
institutes. The most gifted scientists often get crowded out—indeed this is a good
example of spatial barriers reducing competition, allowing survival of a less fit variant.) Yet there is perhaps another factor. Humans produce few progeny, spending
a large proportion of their lives bringing them to adulthood, certainly beyond pu-
berty. The death of a child is unbearable. Massacres, pogroms, mass destruction,
and holocaust leave indelible memories and are impossible to "understand." By
contrast fish produce huge numbers of eggs—the moon fish (*Mola mola*) lays more
than 30 million a time. The point is that the vast majority of progeny never make it
to adulthood. Yet this is an evolutionarily stable strategy. Viruses and microbes fall
into the latter category—holocaust follows each and every infection. What counts
is that ≥ 1 virion or microbe be transmitted from the index case to a susceptible
host. The parent in us focuses more on the potential among his progeny (the interesting mutant), their adaptation to new gardens (niches), turning a blind side to
the virological mayhem.

In an evolutionary sense, all that matters is that a virus maintain a sufficiently
high replicative capacity so as to sustain a chain of transmission. Any weakening
of viral replicative capacity spells disaster. New viruses can and do emerge but
on a scale that is probably billions of times less than the number of viral mutants being generated in the comparable period. This is a profoundly conservative
strategy—merely trying to replicate more than experiment. Ah, that word conservative, anathema to the baby-boomers, comes to the fore. Can it really be so? What
a stultifying picture, in contrast to the shock horror of tabloid newspaper virology,
Preston's *The Hot Zone* and that wonderfully atmospheric, trendy, yet profoundly
ambiguous term, Emerging Viruses.

Conservative perhaps, but is there any suggestion that viruses are more or less so than other replicons? Like extrapolation, choosing examples is frequently dangerous. However let's consider one example, one juxtaposition of data, if only to get synapses firing. We'll choose the eukaryotic and retroviral aspartic proteases.[8] The former exist as a monomer with two homologous domains, while the retroviral counterpart functions as a homodimer. Despite these differences the folding patterns are almost identical. Between humans and chickens there is approximately 38% amino acid divergence among typical aspartic proteases. The HIV-1 and HIV-2 proteases differ by a little more, 52%. No one would doubt the considerable differences in design, metabolism, and lifestyle separating us and chickens. On either side of the HIV protease one finds numerous differences: HIV-1 is vpx^-,vpu^+ while HIV-2 is the opposite, i.e., vpx^+,vpu^-; and there are differences in the size and activities of the *tat* gene product; the LTRs are subtly different. Yet both replicate in the same cells *in vivo*, produce the same disease albeit with different kinetics; HIV-2 infection progresses more slowly. If these differences are too substantial for the reader, consider the 28% divergence between the HIV and chimpanzee SIV proteases. These two viruses are isogenic. Pig and human chromosomal aspartic proteases may differ by around 17%, the differences between these two species being, I think (hope), obvious to all. As one is comparing the same degree of variation, the problem of different generation times is largely overcome. Even by this crude example, the AIDS viruses would seem to be more conservative than mammals in their evolution.

This conclusion is even more surprising when it is realized that HIV is fixing mutations at a rate of 10^{-2}-to-10^{-3} per base per year. By contrast, mammals are fixing mutations approximately one million times less rapidly, i.e., 10^{-8}-to-10^{-9} per base per year.[15] However, the generation times of the two are vastly different, ~1 day for HIV and ~25 years for man. Normalizing for this yields a 100–500-fold higher mutation fixation rate per generation for HIV with respect to humans. Amalgamating this with the preceding paragraph we see that HIV is not only evolving qualitatively in a conservative manner, but it is doing so despite a 100-fold greater propensity to accommodate change. Why? Although HIV can change, its human host is effectively invariant in an evolutionary sense. Probably sticking to the niche is all that matters for HIV, the immensity of viral sequence space allows the random walk to continue. It is as though HIV is in evolutionary stasis; a bit like the coelacanth?

In conclusion, it is clear that there are many ways to be an AIDS virus. On a far more important topic, Paul Simon came to the same conclusion when he wrote his song "Fifty Ways to Leave Your Lover." Nike got it wrong.

REFERENCES

1. Bartel, D. P., and J. W. Szostak. "Isolation of New Ribozymes from a Large Pool of Random Sequences." *Science* **261** (1993): 1411–1418.
2. Borrow, P., H. Lewicki, X. Wei, M. S. Horowitz, N. Peffer, H. Meyers, J. A. Nelson, J. E. Gairin, B. H. Hahn, M. B. Oldstone, and G. M. Shaw. "Antiviral Pressure Exerted by HIV-1-Specific Cytotoxic T Lymphocytes (CTLs) During Primary Infection Demonstrated by Rapid Selection of CTL Escape Mutants." *Nature Med.* **3** (1997): 205–211.
3. Cheynier, R., S. Henrichwark, F. Hadida, E. Pelletier, E. Oksenhendler, B. Autran, and S. Wain-Hobson. "HIV and T Cell Expansion in Splenic White Pulps is Accompanied by Infiltration of HIV-Specific Cytotoxic T Lymphocytes." *Cell* **78** (1994): 373–387.
4. de la Torre, J. C., and J. J. Holland. "RNA Virus Quasispecies Can Suppress Vastly Superior Mutant Progeny." *J. Virol.* **64** (1990): 6278–6281.
5. de la Torre, J. C., E. Martinez-Salas, J. Diez, A. Villaverde, F. Gebauer, E. Rocha, M. Davila, and E. Domingo. "Coevolution of Cells and Virus in a Persistent Infection of Foot-and-Mouth Disease Virus in Cell Culture." *J. Virol.* **60** (1988): 2050–2058.
6. Domingo, E. "RNA Virus Evolution and the Control of Viral Disease." *Prog. Drug Res.* **33** (1989): 93–133.
7. Domingo, E., and J. J. Holland. "Complications of RNA Heterogeneity for the Engineering of Virus Vaccines and Antiviral Agents." In *Genetic Engineering, Principles, and Methods*, edited by J. K. Setlow, 13–31. New York: Plenum Press, 1992.
8. Doolittle, R. F., D. F. Feng, M. S. Johnson, and M. A. McClure. "Origins and Evolutionary Relationships of Retroviruses." *Quart. Rev. Biol.* **64** (1989): 1–30.
9. Drake, J. W. "A Constant Rate of Spontaneous Mutation in DNA-Based Microbes." *Proc. Natl. Acad. Sci., USA* **88** (1991): 7160–7164.
10. Drake, J. W. "Rates of Spontaneous Mutations Among RNA Viruses." *Proc. Natl. Acad. Sci., USA* **90** (1993): 4171–4175.
11. Duarte, E., D. Clarke, A. Moya, E. Domingo, and J. J. Holland. "Rapid Fitness Losses in Mammalian RNA Virus Clones Due to Muller's Ratchet." *Proc. Natl. Acad. Sci., USA* **89** (1992): 6015–6019.
12. Duarte, E. A., I. S. Novella, S. C. Weaver, E. Domingo, S. Wain-Hobson, D. K. Clarke, A. Moya, S. F. Elena, J. C. de la Torre, and J. J. Holland. "RNA Virus Quasispecies: Significance for Viral Disease and Epidemiology." *Infect. Agents Dis.* **3** (1994): 201–214.
13. Eigen, M. "The Viral Quasispecies." *Sci. Am.* **269** (1993): 42–49.
14. Ekland, E. H., J. W. Szostak, and D. P. Bartel. "Structurally Complex and Highly Active RNA Ligases Derived from Random RNA Sequences." *Science* **269** (1995): 364–370.

15. Gojobori, T., and S. Yokoyama. "Molecular Evolutionary Rates of Oncogenes." *J. Mol. Evol.* **26** (1987): 148–156.

16. Ho, D. D., A. U. Neumann, A. S. Perelson, W. Chen, J. M. Leonard, and M. Markowitz. "Rapid Turnover of Plasma Virions and CD4 Lymphocytes in HIV-1 Infection." *Nature* **373** 123–126.

17. Holland, J., K. Spindler, F. Horodyski, E. Grabau, S. Nichol, and X. Van de Pol. "Rapid Evolution of RNA Genomes." *Science* **215** (1982): 1577–1585.

18. Holland, J. J. March meeting on Viral Quasispecies. Santa Fe Institute, Santa Fe, NM, 1996.

19. Holland, J. J., J. C. de le Torre, and D. A. Steinhauer. "RNA Virus Populations as Quasispecies." *Curr. Top. Microbiol. & Immunol.* **176** (1992): 1–20.

20. Huynen, M. A., P. F. Stadler, and W. Fontana. "Smoothness Within Ruggedness: The Role of Neutrality in Adaptation." *Proc. Natl. Acad. Sci., USA* **93** (1996): 397–401.

21. Johnson, P. R., T. E. Hamm, S. Goldstein, S. Kitov, and V. M. Hirsch. "The Genetic Fate of Molecularly Cloned Simian Immunodeficiency Virus in Experimentally Infected Macaques." *Virol.* **185** (1991): 217–228.

22. Kestler, H. W. I., D. J. Ringler, K. Mori, D. L. Panicali, P. K. Sehgal, M. D. Daniel, and R. C. Desrosiers. "Importance of the nef Gene for Maintenance of High Virus Load and for Development of AIDS." *Cell* **65** (1991): 651–662.

23. Maynard Smith, J. "Natural Selection and the Concept of a Protein Space." *Nature* **225** (1970): 563–564.

24. Myers, G., and B. Korber. "The Future of Human Immunodeficiency Virus." In *The Evolutionary Biology of Viruses*, edited by S. S. Morse. New York: Raven Press, 1994.

25. Novella, I. S., S. F. Elena, A. Moya, E. Domingo, and J. J. Holland. "Exponential Increases of RNA Virus Fitness During Large Population Transmissions." *Proc. Natl. Acad. Sci., USA* **92** (1995): 5841–5844.

26. Novella, I. S., S. F. Elena, A. Moya, E. Domingo, and J. J. Holland. "Size of Bottleneck Leading to Fitness Loss is Determined by the Mean Initial Population Fitness." *J. Virol.* **69** (1995): 2869–2872.

27. Olins, P. O., S. C. Bauer, S. Bradford-Golderg, K. Sterbenz, J. O. Polazzi, M. H. Caparon, B. K. Klein, A. M. Easton, K. Paik, and J. A. Klover. "Saturation Mutagenesis of Human Interleukin-3." *J. Biol. Chem.* **270** (1995): 23754–23760.

28. Pelletier, E., W. Saurin, R. Cheynier, N. L. Letvin, and S. Wain-Hobson. "The Tempo and Mode of SIV Quasispecies Development *In Vivo* Calls for Massive Viral Replication and Clearance." *Virol.* **208** (1995): 644–652.

29. Perelson, A. S., A. U. Neumann, M. Markowitz, J. M. Leonard, and D. D. Ho. "HIV-1 Dynamics *In Vivo*: Virion Clearance Rate, Infected Cell Life-Span, and Viral Generation Time." *Science* **271** (1996): 1582–1586.

30. Plikat, U., K. Nieselt-Struwe, and A. Meyerhans. "Genetic Drift Can Dominate Short-Term Human Imunodeficiency Virus Type 1 nef Quasispecies Evolution *In Vivo*." *J. Virol.* **71** (1997): 4233–4240.

31. Quer, J., R. Huerta, I. S. Novella, L. Tsimring, E. Domingo, and J. J. Holland. "Reproducible Nonlinear Population Dynamics and Critical Points During Replicative Competitions of RNA Virus Quasispecies." *J. Mol. Biol.* **264** (1997): 465–471.
32. Schuster, P., P. F. Stadler, and A. Renner. "RNA Structures and Folding. From Conventional to New Issues in Structure Predictions." *Curr. Opin. Struct. Biol.* **7** (1997): 229–235.
33. Sova, P., M. van Ranst, P. Gupta, R. Balachandran, W. Chao, S. Itescu, G. McKinley, and D. J. Volsky. "Conservation of an Intact Human Immunodeficiency Virus Type 1 vif Gene *In Vitro* and *In Vivo*." *J. Virol.* **69** (1995): 2557–2564.
34. Wei, X., S. K. Ghosh, M. E. Taylor, V. A. Johnson, E. A. Emini, P. Deutsch, J. D. Lifson, S. Bonhoeffer, M. A. Nowak, B. H. Hahn, M. S. Saag, and G. M. Shaw. "Viral Dynamics in Human Immunodeficiency Virus Type 1 Infection." *Nature* **373** (1995): 117–122.

Alan D. Frankel† and Andrew D. Ellington‡
†Department of Biochemistry and Biophysics, University of California, San Francisco, CA 94143-0448
‡Department of Chemistry, Indiana University, Bloomington, IN 47405-4001

Combinatorial Approaches to Inhibiting the HIV Rev-RRE Interaction

INTRODUCTION

The human immunodeficiency virus (HIV) Rev protein is essential for viral replication and is, therefore, an attractive target for therapeutic intervention. Rev binds to a specific RNA site, the Rev response element (RRE), located within the *env* region of the viral mRNAs and facilitates the nuclear to cytoplasmic transport of unspliced RNAs that encode the viral structural proteins (see Cullen and also Felber, this volume). The interaction of Rev with the RRE is essential for Rev function and, thus, it may be expected that compounds that target either the protein or RNA will inhibit HIV replication. Here we describe combinatorial strategies that seek to identify peptides or nucleic acid ligands that bind tightly to Rev or to the RRE and block formation of functional RNA-protein complexes. It is hoped that such molecules will serve as templates for further drug design and will increase our understanding of RNA-protein recognition in general.

The Rev protein is 116 amino acids in length and contains several functionally defined domains (Figure 1). The RNA-binding domain is composed of an arginine-rich motif (ARM) of 17 amino acids (residues 34–50) that binds specifically to the

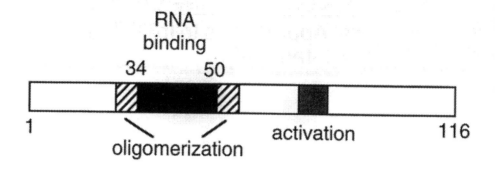

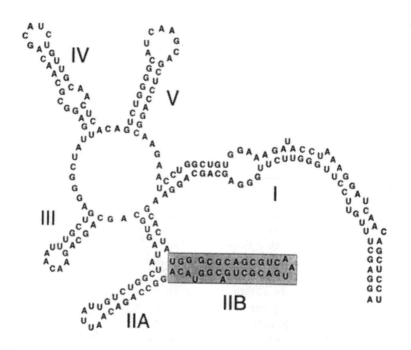

FIGURE 1   Schematic of the Rev protein and its functional domains, and secondary structure of the Rev response element (RRE) with the high-affinity Rev binding site (stem-loop IIB) highlighted.

RRE as an isolated peptide.[23] This domain also serves as the nuclear localization signal for the protein and may help to inhibit splicing of the viral mRNA. Surrounding the RNA-binding domain are amino acids involved in oligomerization.[30] Multiple Rev molecules must bind to the RRE site to function, and these amino acids appear to facilitate multimer formation. Near the C-terminus of the protein is the activation or effector domain (residues 75–83), which is required for nuclear to cytoplasmic transport and appears to physically bridge the Rev-RNA complex to cellular transport proteins.[4,6,7,21]

The RRE is a large structured RNA element (~250 nucleotides) composed of several RNA stem-loops (Figure 1). Within this large element is a high-affinity site (stem-loop IIB) that binds a single Rev monomer, and secondary sites that appear to cooperatively bind additional monomers largely through protein-protein interactions.[32] Disruption of the high-affinity interaction appears sufficient to block Rev function, and the strategies described here focus on possible ways to inhibit this interaction.

DETAILS OF THE HIGH-AFFINITY REV-RRE INTERACTION

There is a substantial amount known about how the arginine-rich RNA-binding domain of Rev recognizes the RRE IIB hairpin, and this knowledge serves as a starting point for the design of inhibitor molecules. The RNA-binding domain and IIB hairpin are shown in Figure 2, highlighting the amino acids and nucleotides known to be important for binding specificity. The peptide adopts an α-helical conformation, which is required for specific RRE binding, and primarily uses six amino acids (four arginines, one threonine, and one asparagine) for recognition.[23] The RNA site is formed largely by an asymmetric internal loop in which two purine-purine base pairs (G:G and G:A) and a bulged (unpaired) nucleotide are used to widen the major groove, allowing access to the helical peptide.[3,18] A standard Watson-Crick RNA helix adopts an A-form geometry with a characteristic narrow and deep major groove, unlike a B-form DNA helix which has a wide and shallow major groove that is easily accessible to a protein. Studies of the Rev-RRE complex and other RNA-protein interactions suggest that bulge nucleotides and non-Watson-Crick base pairs are generally used to widen the RNA major groove and allow recognition of the rich array of functional groups available on bases in the major groove.[26]

Access to a widened major groove not only enhances interactions between Rev and its natural RNA ligands, but may be an inherent requirement for productive Rev-RNA interactions. Rev-binding RNA ligands have been selected from partially or completely random-sequence nucleic acid pools.[1,8,25] Many of the selected binding species (aptamers) resemble the wild-type RRE, and reiterate the importance of the sequences and structures that facilitate the widening of the major groove. For example, sequence covariations found in selected molecules suggest that an A:A homopurine pairing between positions 48 and 71 may present the major groove as well as or better than the wild-type G:G pairing. Additional covariations between

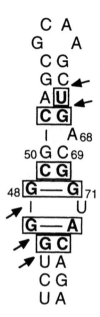

TR_{QA}RRN_{RRR}RWRERQR

FIGURE 2   Nucleotides in stem-loop IIB important for Rev binding are boxed and important phosphates are indicated by arrows, as determined by mutagenesis and chemical modification and substitution experiments. Amino acids in the Rev RNA-binding peptide important for binding are highlighted.

residues 50, 68, and 69 have assisted in the construction of molecular models for stem-loop IIB of the RRE and anti-Rev aptamers, and seem to serve as a molecular "hinge" that can generally assist in widening the major groove.[15]

Recently, structures of the Rev peptide bound to IIB RNA and to an RNA aptamer have been solved by NMR.[2,28] One particularly striking feature in both complexes is the depth of peptide penetration into the major groove (Figure 3). The RNA almost entirely surrounds the α-helix, explaining how six amino acids distributed around the circumference of an α-helix can all contact the RNA. While the NMR data are not sufficient to completely define all the specific amino acid-RNA contacts, many of the contacts can be reasonably inferred in combination with mutagenesis and chemical modification data. As observed in other RNA-protein and DNA-protein complexes, arginine side chains are particularly important for recognizing guanine bases; which they do by forming pairs of hydrogen bonds to the O6 and N7 groups. Some arginines also appear to contact phosphate groups, whose orientation is determined by the local conformation of the RNA backbone. The asparagine side chain hydrogen bonds to groups on the G:A base pair, and the threonine contacts the RNA backbone and also forms an "*N*-cap" that apparently stabilizes the α-helix. Interestingly, the G:G base pair is not directly contacted by the peptide but rather appears to serve a structural role by widening the major groove and helping to define the conformation of the RNA site. As predicted, other residues and substructures also assist in widening the major groove. The anti-Rev aptamer contains an intramolecular base triple in place of the molecular "hinge" formed by residues 50, 68, and 69. This base triple resembles a similar structure found in the TAR element of HIV-1, and in both instances provides access to an infiltrating arginine residue.

As observed in many types of macromolecular interactions, conformational changes are an important feature of the Rev-RRE complex. In this case, both partners of the complex, the peptide and RNA, are relatively disordered on their own and lock into a single conformation upon binding.[3,24] Given that the peptide binds

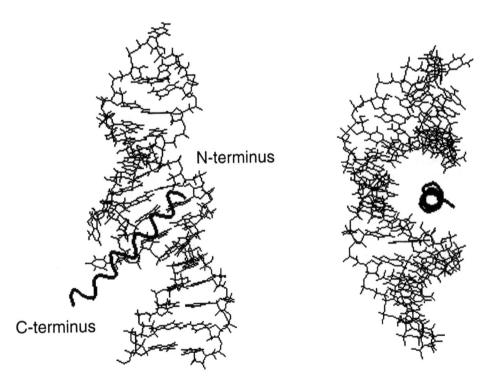

FIGURE 3 Structure of the Rev RNA-binding peptide complexed to stem-loop IIB, determined by NMR. Left panel shows the orientation of the peptide α-helix in the widened major groove, and right panel shows a view down the α-helix axis highlighting the depth of penetration in the groove.

so deeply within the major groove and is largely surrounded by the RNA, it seems reasonable that a conformational change in the RNA is needed to allow formation of such an extensive binding interface. Biochemical experiments indicate that the peptide α-helix must be at least partially formed prior to binding in order to specifically recognize the RRE site, but the helix is rather unstable when isolated from the rest of the Rev protein and becomes fully helical only upon RNA binding.[24] Again, this biochemical feature appears to be general: RNA molecules selected to bind to the peptide do not resemble the RRE, but can also induce conformational changes.[27] It is not known whether the RNA-binding helix is stable in the context of the complete protein framework or whether an induced fit mechanism accompanies protein binding as well. The arrangement of Rev subunits within the oligomeric Rev-RRE complex is not known, though the high-affinity IIB interaction appears to nucleate binding of additional subunits to adjacent regions of the RRE.[32]

IDENTIFICATION OF RNA-BASED INHIBITORS BY *IN VITRO* SELECTION

While *in vitro* selection provides an excellent tool for identifying sequence and structural elements that may be important for Rev-RNA interactions, selected nucleic acid sequences may also be useful as potential therapeutics. Aptamers selected from a completely random sequence pool bound Rev from 3-fold to 10-fold better than stem-loop IIB of the RRE.[8,25] When the aptamers were substituted into the RRE in place of IIB they successfully facilitated the Rev-dependent transport of viral mRNAs.[22] These results suggest that stem-loop IIB serves as a passive platform for the accumulation of Rev, and that Rev-dependent mRNA transport is not dependent on a particular set of interactions or RNA-induced conformational changes. Interestingly, while all of the aptamers appeared to function better than the wild-type, no correlation was observed between *in vitro* binding affinity and *in vivo* mRNA transport activity.

Antisense oligonucleotides,[14] or catalytic RNAs[29] directed against Rev or the RRE have shown promise as therapeutic reagents. Therefore, it should also prove possible to disrupt Rev-RRE interactions by developing RNA decoys that target Rev. When the high-affinity Rev-binding site, IIB, is expressed in tissue culture cells it provides some protection against HIV-1 infection and replication.[16] Since anti-Rev aptamers seem to function better than IIB in tissue culture cells, it was hoped they might also prove to be better RNA decoys for Rev. Aptamer sequences have been inserted into RNA expression vectors behind strong pol III promoters and co-transfected into tissue culture cells along with HIV-1 proviral DNA. Under these experimental conditions, 20-fold lower levels of HIV-1 progeny were produced relative to "no aptamer" controls.[9] However, the aptamers are no more efficacious than wild-type IIB.

It is possible that aptamers may also be used as conventional pharmaceuticals rather than gene therapy reagents. However, RNA molecules are extremely unstable in human sera and have half-lives of mere minutes. Nucleotides containing modifications at the 2' position have been shown to greatly increase nucleic acid stability against nucleases. For example, 2'-amino-cytidine and 2'-amino-uridine can be introduced into nucleic acids by polymerases and have been introduced into selection experiments.[12] Selected, modified aptamers have been shown to be stable in urine or serum for up to a day.[10] We have now selected anti-Rev aptamers from modified RNA pools and have found that they are similarly stable in serum. Such aptamers can be shortened and further stabilized by introducing nonnucleotide cross-linkers into helices. A trans-stilbene bridge has been incorporated into stem-loop IIB and anti-Rev aptamers; the resultant 22-nucleotide compounds could bind Rev almost as well as a 94-nucleotide fragment of the RRE.[17] Finally, circular forms of IIB have been generated using the Anabaena pre-tRNA group I intron self-splicing permuted intron-exon (PIE).[19] These circular RNAs are much more resistant to exonuclease degradation.

The efficacy of anti-Rev aptamers can also be increased by including modified bases during selection experiments. Five-iodouridine has been used to produce photocrosslinks between modified RNA molecules and proteins, and has also been introduced into selection experiments that target Rev.[13] Ligands were identified that were able to crosslink to Rev in the presence of UV radiation; the crosslinks could specifically form even in cell extracts.

IDENTIFICATION OF PEPTIDE-BASED INHIBITORS BY *IN VIVO* SELECTION

A complementary approach to selecting RNAs that target Rev is to select peptides or other molecules that target the RRE. An *in vivo* system based on transcription antitermination by the bacteriophage λN protein has been developed to identify RRE-binding peptides.[11] In this system, the *nut* site to which N usually binds is replaced by the RRE IIB hairpin and a peptide library is fused to the amino-terminus of N (Figure 4). If a peptide in the library binds to the RRE, the fused N protein is assembled into a functional antitermination complex, resulting in β-galactosidase (lac Z) expression and formation of a blue colony. With this assay, several high affinity RRE-binding peptides were identified from a library in which only arginine, serine, and glycine were allowed at nine randomized positions of an

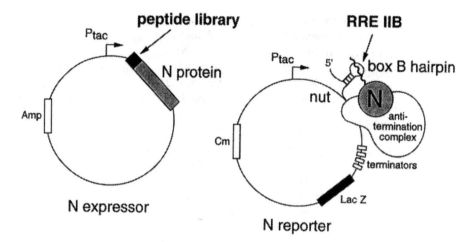

FIGURE 4 The two-plasmid system for measuring transcription antitermination by the lambda N protein. Peptides are fused to the amino terminus of N and are expressed from a tac promoter. β-galactosidase expression from the reporter plasmid requires binding to RRE IIB, mediated by the fused peptide. RRE IIB replaces the box B hairpin of the nut site, which provides the binding site for wild-type N.

otherwise arginine-rich peptide. For one peptide studied in detail, binding to the RRE occurred with slightly higher affinity and specificity than the Rev peptide itself but unlike Rev, circular dichroism spectroscopy indicated no α-helix formation. Thus, there may be multiple structural solutions to recognition of the RRE IIB site. The peptide has been subequently mutagenized and even tighter binders have been evolved after two additional rounds of selection. These tight RRE-binding peptides displace Rev from the RRE *in vitro*, correlating with their binding affinities, and inhibit Rev activity in tissue culture assays. Thus, it seems possible to block the Rev-RRE interaction and inhibit Rev function by specifically targeting either the protein or RNA. Several small molecules, including neomycin and other antibiotics, also have been found to target the RRE, block Rev binding, and inhibit Rev function.[20,31]

PROSPECTS FOR THE FUTURE

The strategies described here may be viewed as starting points in the drug discovery process. RNAs and peptides specifically targeted toward the Rev protein or RRE can effectively block the Rev-RRE interaction, although it is still unclear whether these types of macromolecules will themselves be therapeutically useful. Instead, as more detailed structural and structure/function information is obtained, it may be possible to design small molecule analogs that could be further developed into drugs. Conversely, selection approaches can help to define the specificities of more conventional drugs by identifying which sequences interact with which compounds.[5] RNAs or peptides expressed intracellularly may also be evaluated for gene therapy approaches. The essential nature of the Rev-RRE interaction for HIV replication makes it an attractive therapeutic target.

Fundamental questions remain about the Rev-RRE interaction and RNA-protein interactions in general. Combinatorial approaches such as those described here may be expected to shed light on some of these questions. For example, what types of RNA structures can be used to recognize protein targets? What types of peptide structures and amino acid side chains can be used to recognize RNA? What strategies do proteins use to recognize different types of RNA structures, including duplexes, internal loops (as in the RRE), terminal loops, and bulges? How do the affinities and specificities of selected RNAs and peptides compare to natural RNAs and proteins? As sequence space is more fully explored, one may anticipate some interesting and unexpected answers.

REFERENCES

1. Bartel, D. P., M. L. Zapp, M. R. Green, and J. W. Szostak. "HIV-1 Rev Regulation Involves Recognition of Non-Watson-Crick Base Pairs in Viral RNA." *Cell* **67** (1991): 529–536.
2. Battiste, J. L., H. Mao, N. S. Rao, R. Tan, D. R. Muhandiram, L. E. Kay, A. D. Frankel, and J. R. Williamson. "Alpha Helix Major Groove Recognition in an HIV-1 Rev Peptide-RRE RNA Complex." *Science* **273** (1996): 1547–1551.
3. Battiste, J. L., R. Tan, A. D. Frankel, and J. R. Williamson. "Binding of an HIV Rev Peptide to Rev Responsive Element RNA Induces Formation of Purine-Purine Base Pairs." *Biochemistry* **33** (1994): 2741–2747.
4. Bogerd, H. P., R. A. Fridell, S. Madore, and B. R. Cullen. "Identification of a Novel Cellular Cofactor for the Rev/Rex Class of Retroviral Regulatory Proteins." *Cell* **82** (1995): 485–494.
5. Ellington, A. D. "Using In Vitro Selection for Conventional Drug Design." *Drug Devel. Res.* **33** (1994): 102–115.
6. Fischer, U., J. Huber, W. C. Boelens, I. W. Mattaj, and R. Luhrmann. "The HIV-1 Rev Activation Domain is a Nuclear Export Signal that Accesses an Export Pathway Used by Specific Cellular RNAs." *Cell* **82** (1995): 475–483.
7. Fritz, C. C., M. L. Zapp, and M. R. Green. "A Human Nucleoporin-Like Protein that Specifically Interacts with HIV Rev." *Nature* **376** (1995): 530–533.
8. Giver, L., D. Bartel, M. Zapp, A. Pawul, M. Green, and A. D. Ellington. "Selective Optimization of the Rev-Binding Element of HIV-1." *Nucleic Acids Res.* **21** (1993): 5509–5516.
9. Good, P. D., A. J. Krikos, S. X. L. Li, E. Bertrand, N. S. Lee, L. Giver, A. D. Ellington, J. A. Zaia, J. J. Rossi, and D. R. Engelke. "Expression of Small, Therapeutic RNAs in Human Cell Nuclei." *Gene Therapy* **4** (1997): 45–54.
10. Green, L., D. Jellinek, C. Bell, L. A. Beebe, B. D. Feistner, S. C. Gill, F. M. Jucker, and N. Janjic. "Nuclease-Resistant Nucleic Acid Ligands to Vascular Permeability Factor/Vascular Endothelial Growth Factor." *Chem. Biol.* **2** (1995): 683–695.
11. Harada, K., S. S. Martin, and A. D. Frankel. "Selection of RNA-Binding Peptides In Vivo." *Nature* **380** (1996): 175–179.
12. Jellinek, D., L. S. Green, C. Bell, C. K. Lynott, N. Gill, C. Vargeese, G. Kirschenheuter, D. P. C. McGee, P. Abesinghe, and W. A. Pieken. "Potent 2'-Amino-2'-Deoxypyrimidine RNA Inhibitors of Basic Fibroblast Growth Factor." *Biochemistry* **34** (1995): 11363–11372.
13. Jensen, K. B., B. L. Atkinson, M. C. Willis, T. H. Koch, and L. Gold. "Using In Vitro Selection to Direct the Covalent Attachment of Human Immunodeficiency Virus Type 1 Rev Protein to High-Affinity RNA Ligands." *Proc. Natl. Acad. Sci., USA* **92** (1995): 12220–12224.

14. Kim, J. H., R. J. McLinden, J. D. Mosca, M. T. Vahey, W. C. Greene, and R. R. Redfield. "Inhibition of HIV Replication by Sense and Antisense Rev Response Elements in HIV-Based Retroviral Vectors." *J. Acq. Imm. Def. Syndr. Hum. Retrovirol.* **12** (1996): 343–351.

15. Leclerc, F., R. Cedergren, and A. D. Ellington. "A Three-Dimensional Model of the Rev-Binding Element of HIV-1 Derived from Analyses of Aptamers." *Nat. Struct. Biol.* **1** (1994): 301–310.

16. Lee, S. W., H. F. Gallardo, E. Gilboa, and C. Smith. "Inhibition of Human Immunodeficiency Virus Type 1 in Human T Cells by a Potent Rev Response Element Decoy Consisting of the 13-Nucleotide Minimal Rev-Binding Domain." *J. Virol.* **68** (1994): 8254–8264.

17. Nelson, J. S., L. Giver, A. D. Ellington, and R. L. Letsinger. "Incorporation of a Non-Nucleotide Bridge into Hairpin Oligonucleotides Capable of High-Affinity Binding to the Rev Protein of HIV-1." *Biochemistry* **35** (1996): 5339–5344.

18. Peterson, R. D., D. P. Bartel, J. W. Szostak, S. J. Horvath, and J. Feigon. "1H NMR Studies of the High-Affinity Rev Binding Site of the Rev Responsive Element of HIV-1 mRNA: Base Pairing in the Core Binding Element." *Biochemistry* **33** (1994): 5357–5366.

19. Puttaraju, M., and M. D. Been. "Generation of Nuclease Resistant Circular RNA Decoys for HIV-Tat and HIV-Rev by Autocatalytic Splicing." *Nucl. Acids Symp. Ser.* **33** (1995): 152–155.

20. Ratmeyer, L., M. L. Zapp, M. R. Green, R. Vinayak, A. Kumar, D. W. Boykin, and W. D. Wilson. "Inhibition of HIV-1 Rev-RRE Interaction by Diphenylfuran Derivatives." *Biochemistry* **35** (1996): 13689–13696.

21. Stutz, F., M. Neville, and M. Rosbash. "Identification of a Novel Nuclear Pore-Associated Protein as a Functional Target of the HIV-1 Rev Protein in Yeast." *Cell* **82** (1995): 495–506.

22. Symensma, T. L., L. Giver, M. Zapp, G. B. Takle, and A. D. Ellington. "RNA Aptamers Selected to Bind Human Immunodeficiency Virus Type 1 Rev In Vitro are Rev Responsive In Vivo." *J. Virol.* **70** (1996): 179–187.

23. Tan, R., L. Chen, J. A. Buettner, D. Hudson, and A. D. Frankel. "RNA Recognition by an Isolated Alpha Helix." *Cell* **73** (1993): 1031–1040.

24. Tan, R., and A. D. Frankel. "Costabilization of Peptide and RNA Structure in an HIV Rev Peptide-RRE Complex." *Biochemistry* **33** (1994): 14579–14585.

25. Tuerk, C., and S. MacDougal-Waugh. "In Vitro Evolution of Functional Nucleic Acids: High-Affinity RNA Ligands of HIV-1 Proteins." *Gene* **137** (1993): 33–39.

26. Weeks, K. M., and D. M. Crothers. "Major Groove Accessibility of RNA." *Science* **261** (1993): 1574–1577.

27. Xu, W., and A. D. Ellington. "Anti-Peptide Aptamers Recognize Amino Acid Sequence and Bind a Protein Epitope." *Proc. Natl. Acad. Sci., USA* **93** (1996): 7475–7480.

28. Ye, X., A. Gorin, A. D. Ellington, and D. J. Patel. "Deep Penetration of an Alpha-Helix into a Widened RNA Major Groove in the HIV-1 Rev Peptide-RNA Aptamer Complex." *Nat. Struc. Biol.* **3** (1996): 1026–1033.

29. Yuyama, N., J. Ohkawa, T. Koguma, M. Shirai, and K. Taira. "A Multifunctional Expression Vector for an Anti-HIV-1 Ribozyme that Produces a 5'- and 3'-Trimmed Trans-Acting Ribozyme, Targeted Against HIV-1 RNA, and *Cis*-Acting Ribozymes that Are Designed to Bind to and Thereby Sequester Trans-Activator Proteins such as Tat and Rev." *Nucl. Acids Res.* **22** (1994): 5060–5067.

30. Zapp, M. L., T. J. Hope, T. G. Parslow, and M. R. Green. "Oligomerization and RNA Binding Domains of the Type 1 Human Immunodeficiency Virus Rev Protein: A Dual Function for an Arginine-Rich Binding Motif." *Proc. Natl. Acad. Sci., USA* **88** (1991): 7734–7738.

31. Zapp, M. L., S. Stern, and M. R. Green. "Small Molecules that Selectively Block RNA Binding of HIV-1 Rev Protein Inhibit Rev Function and Viral Production." *Cell* **74** (1993): 969–978.

32. Zemmel, R. W., A. C. Kelley, J. Karn, and P. J. Butler. "Flexible Regions of RNA Structure Facilitate Co-operative Rev Assembly on the Rev-Response Element." *J. Mol. Biol.* **258** (1996): 763–777.

Roland Somogyi
National Institutes of Health, Laboratory of Neurophysiology, NINDS, Building 36/2C02,
Bethesda, MD 20892; E-mail: rolands@helix.nih.gov

States, Trajectories, and Attractors: A Genetic Networks Perspective of Viral Pathogenesis

INTRODUCTION

Advances in molecular biological research have enabled the identification of the human immunodeficiency virus (HIV) and the sequencing of its genome. Much is now known about viral structures, but many facets of HIV pathogenesis still elude us. Although the virus carries relatively little genetic information, its interactions with the host are complex and reach beyond a few simple molecular targets on a single cell type. Understanding such complexity may be aided by the concept of the genetic network. We now know that cellular signaling and the regulation of gene expression are processes that involve the products of many genes, often in intriguing combinatorial interactions. Within such a complex regulatory web, macroscopic function may in turn be distributed across several genes, while the function of a particular gene may be distributed over several macroscopic features (Figure 1). In no way is causality violated in such complex systems, but fundamental understanding will require conceptualization within parallel distributed processing networks. Most fundamentally, the expression of genes activates protein functions (strictly determined by gene structure) that in turn regulate the expression of more genes. Implicitly, genetic networks encompass all levels of signaling and could serve as a universal model for the organism's regulatory network.[8]

With respect to HIV pathogenesis, we have observed changes in the expression of a wide range of host genes in various cell types.[7] These changes may be important in the etiology of AIDS and should provide clues to pharmacological targeting. A better understanding of the host's genetic network architecture and a close characterization of the perturbations caused by the virus may lead to the identification of critical groups of molecules that underlie various AIDS symptoms. The experimental tools for large-scale gene expression mapping[11] are now becoming available and should allow for a general mapping of virally induced changes in host gene expression. Observing these changes over time may give us clues to causal links between viral and host genes, beyond the most direct interactions. In parallel, computational techniques (a) are now established for studying the dynamic principles of genetic networks in Boolean network models, and (b) are being developed for the analysis of large-scale gene expression data and the heuristic extraction of biological genetic network architectures. Following such an interdisciplinary strategy will bring us closer to understanding the intricacies of viral perturbations of the host system. This will be critical to the design of the combinatorial therapeutic strategies necessary to "outsmart" and systematically block the viral invaders.

A GENETIC NETWORKS PERSPECTIVE OF VIRAL PATHOGENESIS

Viruses that carry very little genetic information have learned to manipulate the genetic information of higher organisms on a fundamental level. The virus can make multiple hits, thereby affecting the expression of a wide variety of genes (see Table 1)—in the case of HIV this is not limited to lymphocytes.[2] How can such a simple structure produce such a complex response?

Understanding the interactions of a relatively simple viral system with a complex host system requires the introduction of a genetic network. The effect of viral genes must be understood in terms of the context of their target cell. Depending on the state of the host, i.e., which gene targets are expressed and how these targets interact with other gene products, expression of viral genes may lead to a variety of pleiotropic effects. All of this information can be processed within a genetic network.

TABLE 1 Pleiotropic effects of viral genes in the perturbation of host gene expression (based on Rosenblatt et al.[7]).

HIV	
•	integration into host genome
•	expresses 20 different mRNAs
•	produces pathological symptoms beyond those associated with T-cell infection
Targets of HIV TAT gene	
•	TGFβ, TNFβ, IL-6, IL-1, GM-CSF
•	downregulation of: C28, MHC class 1
•	soluble, extracellular TAT: competes with vitronectin binding
•	fibronectin, collagens 1 and 3
•	neurotoxic effects, connections with NOS and NMDA receptors, prion genes
•	oncogenes c-fos and c-fms (alteration of immune cell differentiation)
HTLV	
•	T-cell lymphonas (lack of specific proviral integration!, transformation through TAX— analogous to a state change leading to an alternative attractor)
•	associated with chronic neurological diseases
Targets of HTLV TAX gene	
•	interaction with host transacting of elements
•	e.g., IL-2, IL2 receptor α chain (autocrine loop), c-myc, c-fos, EGR-1 synergism with NFκB—all relevant to T-cell transformation
•	PTHrP (parathyroid hormone-related protein)— relevant to clinically observed hypercalcemia

PRINCIPLES OF GENETIC NETWORKS

We are faced with the observation that we cannot reduce living systems to a simple one-to-one mapping of a particular molecular parameter to a single particular function or even to another single molecular parameter (Figure 1). However, it is still useful to reduce biomolecules to genes and their products since gene products encompass a large number of the measurable biomolecules that carry a major portion of the information in the system. A particular function, or the activity of a particular gene, generally depends on the activity of several genes, i.e., is subject to multigenic regulation. By the same token, a single gene can interact with and regulate a variety of downstream genes or functions (pleiotropic regulation, as particularly relevant to virus-host interactions). Together, this cross-wiring of gene function forms a genetic network. While we cannot deny the fact of the genetic network, its implied complexity may appear overwhelming to many researchers. But there may be a way to understand the principles of these dynamics in genetic networks that will lead to the discovery of applications for characterizing, predicting, and manipulating these systems.

Using a few simplifying assumptions, we can construct Boolean network models of genetic networks to help us understand the nature of a parallel distributed processing biomolecular system.[3] Presuming that highly sigmoid interactions or threshold functions underly the molecular interactions in gene regulation, we assume the limiting case that a gene will be turned on or off depending on whether its regulatory inputs are on or off. If a gene has more than one gene as its regulatory input, we must define the rule, or combinations of the on-and-off state of the inputs, that will map activate or inactivate it. These rules are by definition Boolean, i.e., they can be described as a combination of logical *and* and *and not* functions. Essentially, the state or combination, of on-and-off values of input elements at a time point t will determine whether the output gene will be on or off at time point $t + 1$.

A simple example illustrates this point (for a detailed description, see Somogyi and Sniegoski[8]). In Figure 2 we show a simple, 3-element network, its wiring diagram, and the rules. In this case, C will inhibit the activation of A by

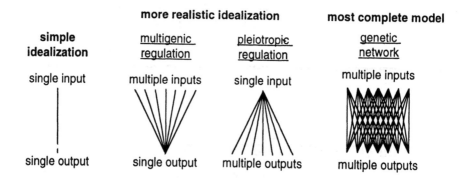

FIGURE 1 Principles of biological networks. Our present knowledge of the molecular processes underlying signaling and the regulation of gene expression teaches us that many processes depend on the combinatorial interactions of a variety of individual effectors. Together these form a parallel, distributed processing feedback system, the genetic network.

| wiring and rules | trajectory 1 (point attractor) | trajectory 2 (dynamic attractor) |

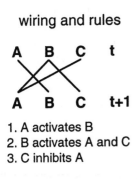

A   B   C   t

A   B   C   t+1

1. A activates B
2. B activates A and C
3. C inhibits A

iteration	A	B	C
1	1	1	0
2	1	1	1
3	0	1	1
4	0	0	1
5	0	0	0
6	0	0	0

iteration	A	B	C
1	1	0	0
2	0	1	0
3	1	0	1
4	0	1	0

FIGURE 2 Trajectory and attractors of a simple Boolean network. The wiring diagram and rules determine the regulatory interactions between the "genes." Each on/off pattern forms a state, which determines the state at the next time point according to the wiring and rules, forming a trajectory, or time series. The upper trajectory results in a permanently repeating all off state. The lower trajectory reaches an infinite 2-state cycle and a dynamic attractor.

B, the rule for A, therefore, being "B and not C." One can easily determine the time series of state changes, or trajectory, by following these rules. Starting from the state $A = 1$, $B = 1$, and $C = 0$ (trajectory 1), we find that after five time steps all elements will be off and continue to be off ad infinitum. This means that the trajectory has reached a repeating state, or point attractor. However, if one calculates the time course from the state $A = 1$, $B = 0$, and $C = 0$, after three time steps the network reaches a 2-state cycle, referred to as a dynamic attractor. Since an $n = 3$ binary network can generate no more than 2^n states, the complete state space of 8 states is captured in trajectories 1 and 2. This type of behavior holds true for all Boolean networks: each state maps to exactly one other state at the next time point, the succession of states in time define a trajectory, which inexorably leads to a point, or dynamic, attractor.

The structures that Boolean networks can form become more interesting with an increasing number of elements. We have constructed a 27-element model[8] according to a hierarchy of "signaling" genes (A1-C1, A3-C3, P-R) that interact with one another to determine the state of "structural genes" (S1-S6, U1-U6, T1-T6), which one could envision as sets of cell-type specific genes (top panel, Figure 3). Since all members of each group of structural genes receive the same inputs and rules, each group can be reduced to one characteristic member, S, U, or T, to determine the trajectory (right panel, Figure 3). After eight time steps, the trajectory reaches a 6-state dynamic attractor. The trajectory serves as an analogy to the gene expression series encountered during development, while the attractor can be likened to a differentiated cell type.[3] Note that while the signaling genes are oscillating, the structural genes do not fluctuate in the attractor. Parenthetically, a general prediction of this type of modeling is that gene expression patterns in stable, differentiated cells may exhibit this cycling behavior outside of known rhythms such as the cell cycle and the circadian clock. This prediction could easily hold true even assuming nonlinear molecular interactions of a steepness short of the extreme on/off function.

It may initially appear surprising that these complex parallel processing networks exhibit a great deal of convergence and stability. This may be counter to our intuition that might lead us to believe that a system depending on the coordination of many interacting parts would by nature be very tricky and unstable. These features are illustrated in Figure 3, where we have plotted the trajectory shown in the upper right corner as a series of connected points labeled with the time step number.

FIGURE 3 Basin of attraction of a Boolean network. All endpoints and nodes of the lines in the central graph represent states. These are connected to each other through time series (represented by the lines), terminating in the attractor. See text and Somogyi and Sniegoski,[8] for a more detailed description. Graphics were generated using the DDLAB software.[13]

The attractor is shown as a 6-state cycle between states 8 to 13. The important point is that, starting at the transition between state 1 and state 2, we see that state 2 can be reached not only from state 1, but from several other states, one at each end point of the line leading to 2. The same applies to all other states of the trajectory. Therefore this graph is called a basin of attraction of the network[12]—all states at the end of each line and at each node connect lines converging to the same attractor cycle. Of course, any random change of a state to another state within this basin will not change the outcome of the network. This confers stability; the network does not depend on one particular trajectory to reach its destination— there are many equal possibilities—meaning it is not easy for the system to go seriously astray. In summary, convergence and stability in Boolean networks are manifested in our model because (a) the network is characterized by a relatively small number of final patterns, i.e., attractors; (b) these attractors can be reached from a variety of starting states; and (c) minor state changes usually result in an alternative trajectory leading to the same attractor.

However, this network has more than one attractor—8 to be exact—each with its own basin of attraction as shown in the basin of attraction field of Figure 4.

FIGURE 4 Complete basin of attraction field of a 12-gene Boolean network. The network covers $2^{12} = 4096$ states. All of these states are represented by endpoints of lines or nodes between lines in basin of attraction graphs. The time course of any state in the network must lead to one of the 8 attractors. The attractors themselves cover 27 states out of the possible 4096, demonstrating the high degree of convergence of this example. Graphics were generated using DDLAB software.[13]

Therefore, it is possible that a random alteration of a state may lead to another state, which is part of another basin of attraction. Such state changes have been used as an analogy for transdifferentiation maps of cell types, i.e., how, by perturbation, cell types can be transformed into one another.[3]

We may, therefore, view disease as a perturbation of the genetic network, leading into pathological attractors. This is represented by cancer and by cells genetically manipulated through the activation of viral genes (schematized in Figure 5). Following a "developmental" transient, the trajectory "differentiates" into a stable dynamic attractor. But following a perturbation, a permanent switching "off" of a gene in S3, the trajectory follows a transient pattern of state changes, leading to the all-off attractor, which can be likened to cell death. Of course, the gene that has been altered in its expression in this example may not be critical for another basin of attraction. Therefore, the significance of a point perturbation by a virus will be highly context dependent (as we observe, not all cell types are affected by the virus). Only viruses that have overcome the redundancy within the network by identifying the vulnerable targets will be recognized as serious pathogens, but those are exactly the ones we are concerned with.

A STRATEGY FOR INTEGRATING GENETIC NETWORKS INTO HIV RESEARCH

We have discussed the pleiotropic effects of viral genes in perturbing the host organism and the principles of genetic network dynamics using the analogy of Boolean networks. How can this perspective be integrated into a practical strategy for understanding and dealing with HIV infection? The first issue in network analysis is carrying out the multiparameter measurements required for defining a state. Since the activity of genes defines the output of the genetic network, the state of the system is reflected in the pattern of expressed mRNAs. Today, we are in the position to make these state measurements using newly developed technologies for large-scale gene expression mapping (GEM). Such assays could be based on automated RT-PCR (reverse transcription-polymerase chain reaction),[11] hybridization chips, Serial Analysis of Gene Expression (SAGE), or alternative technologies which are in development.[6] Another important factor is knowing the gene sequences on which to base the design of various gene-specific hybridization probes. Considering the progress of the genome sequencing projects, much information is already at hand, while several complete genomes, including human, will be completed in the foreseeable future. In essence, we are able to produce state measurements of genetic networks, including a sizable number of genes, and can expect to measure the complete gene expression matrix covering all genes in the not too distant future.

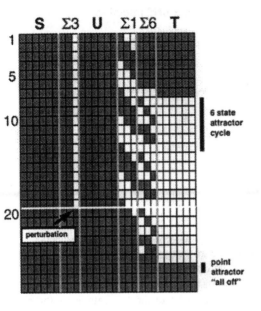

FIGURE 5 Response to perturbation in a genetic network. After reaching a stable 6-state cycle, the network trajectory experiences an "external" perturbation, causing the inactivation of a gene. This leads to a series of state changes, resulting in a new attractor, the all-off state in this case.

As a first step, gene expression mapping of host cells or tissues affected by the virus will provide us with a clear diagnosis of how deeply the host system has been affected. It will capture the pleiotropic effects of viral infection and gene expression, many of which will probably be surprising given a large enough matrix of genes. Of course, these changes will vary depending on the affected tissue and cell types (schematized in Figure 6). The comparison of samples from a sufficient number of animals (e.g., using monkeys as models[5]) or post mortem human tissues, may indicate general sets of perturbed host genes.

As a second step, time courses could be studied in animal models, providing data analogous to the trajectories in Boolean nets. Deeper inference could be drawn from such data regarding cause and effect. In the future we hope to see methods for "reverse engineering" of network architectures from trajectories.[4,9] Of course, having detected the changes in host gene expression, one will be presented with a large set of clues to which processes underlie, e.g., CNS pathology. Integrating these discoveries with knowledge on other gene interactions may result in therapeutic drug strategies focusing on the identified genes. From the network perspective, which explains how function is distributed across several genes, one may look toward combinatorial therapies targeted at several selected genes simultaneously. Early beginnings of such multidrug strategies have already had significant successes so far in AIDS therapy.[1]

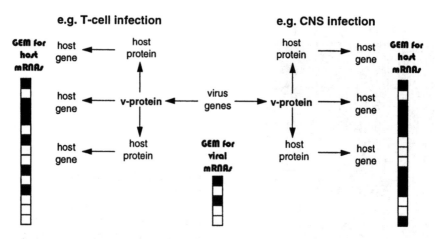

FIGURE 6   Understanding the viral perturbation of host genetic networks through gene expression mapping (GEM). The expression of viral genes, a direct result of infection, will lead to differential responses in the host's genetic network depending on the background gene expression, or context, of the infected host tissue. Mapping these changes will give us insight into effector actions causing the pathology, thereby providing potential therapeutic molecular targets.

REFERENCES

1. James, J. S. "New View of Antiretroviral Treatment: Illustrations from 'Late Breaker' Abstracts." *AIDS Treatment News* **251** (1996): 4–6.
2. Johnson, R. T. "The Pathogenesis of HIV Infections of the Brain." *Curr. Top. Microbiol. Immunol.* **202** (1995): 3–10.
3. Kauffman, S. A. *The Origins of Order: Self-Organization and Selection in Evolution.* Oxford: Oxford University Press, 1993.
4. Liang, S., S. Fuhrman, and R. Somogyi. "REVEAL, A General Reverse Engineering Algorithm for Inference of Genetic Network Architectures." Proceedings of the Pacific Symposium on Biocomputing, 1998.
5. Narayan, O., S. V. Joag, and E. B. Stephens. "Selected Models of HIV-Induced Neurological Disease." *Curr. Top. Microbiol. Immunol.* **202** (1995): 151–166.
6. Nowak, R. "Entering the Postgenome Era." *Science* **270** (1995): 368–369.
7. Rosenblatt, J. D., S. Miles, J. C. Gasson, and D. Prager. "Transactivation of Cellular Genes by Human Retroviruses." *Curr. Top. Microbiol. Immunol.* **193** (1995): 25–49.
8. Somogyi, R., and C. A. Sniegoski. "Modeling the Complexity of Genetic Networks: Understanding Multigenic and Pleiotropic Regulation." *Complexity* **1(6)** (1996): 45–63.
9. Somogyi, R., S. Fuhrman, M. Askenazi, and A. Wuensche. "The Gene Expression Matrix: Towards the Extraction of Genetic Network Architectures." *Proceedings of the World Congress of Nonlinear Analysts*, Elsevier Science, 1996.
10. Somogyi, R., X. Wen, W. Ma, and J. L. Barker. "Developmental Kinetics of GAD Family mRNAs Parallel Neurogenesis in the Rat Spinal Cord." *J. Neuroscience* **15** (1995): 2575–2591.
11. Wen, X., S. Fuhrman, S. Smith, G. S. Michaels, D. B. Carr, J. L. Barker, and R. Somogyi. "Gene Expression Mapping of the Developing Spinal Cord: Identification of Expression Clusters." *Proc. Natl. Acad. Sci., USA*, in press.
12. Wuensche, A. "The Ghost in the Machine: Basins of Attraction in Random Boolean Networks." In *Artificial Life III*, edited by C. G. Langton, 465–501. Santa Fe Institute Studies in the Sciences of Complexity, Proc. Vol. XVII. Reading, MA: Addison-Wesley, 1993.
13. Wuensche, A. "Discrete Dynamics Lab (DDLAB)." Software and documentation at MIT Press Artificial Life Online, 1995. (alife.santafe.edu/alife/software/ddlab.html).

Christine Neuveut and Kuan-Teh Jeang
Molecular Virology Section, National Institute of Allergy and Infectious Diseases, National Institutes of Health, Bethesda, MD 20892–0460

Some Reflections on Complexities in Understanding Virus Biology

Over the past decade and one-half, the biology of HIV-1 has been an intensely investigated area. A search of MEDLINE using the term HIV-1 reveals 34,624 citations in the most recent five-year period. Quantitatively, this is a daunting number; however, qualitatively, much of the published literature can be contradictory and confusing. So much so, that HIV researchers (including ourselves) frequently lament about this seemingly unique predicament. Upon further reflection, it is apparent that HIV is neither uniquely confusing nor uniquely complex. The complexity of HIV is one that is inherent to all biological systems. The actual complexity of virus-cell interactions is further exacerbated by common expectations of simple "big picture" answers despite full awareness of an inability to control for multiple known and unknown variables. Added to this is the myriad of different and frequently nonequivalent experimental systems used to answer the same biological questions. Hence, framed in the context of an analogy, in order to solve definitively for N unknowns in a mathematical equation one should have N or $N + X$, where X is any integer larger than 0, independent equations; virus-cell/host interactions offer a challenge where N is large and largely unknown. Furthermore, N is likely to be of a size that exceeds the number of "constant equations" one has available to use for solution. Thus, strict, and perhaps unwarrranted, expectations of certainty or rigid interpretations of data in settings that are fraught with inherent uncertainties can lead to missteps and confusion. What might be some examples of these difficulties in virus-cell biology? In a personally selective and anectodal fashion, we describe some examples of complex biological interpretations in virus-cell interactions involving HIV and other pathogenic viruses.

HSV, HPV, AND THE CAUSE OF CERVICAL CARCINOMA

Prior to the isolation and characterization of human retroviruses, human herpes viruses (HHV) represented one of the best and most popularly studied viral pathogens (2274 MEDLINE citations for the term herpes simplex virus [HSV or more recently reclassified as HHV1] in the most recent five-year period). Herpes viruses are large DNA viruses, for example, the genome size of cytomegalovirus (CMV) exceeds 200 kbp when by comparison the HIV-1 proviral DNA size is approximately 9 kbp. Thus, the biology of herpes infections is expected to be highly complex. Diseases caused by herpes viruses are significant and varied. CMV is a known and frequent cause of birth defects, and it is the most common virus to induce opportunistic complications in HIV disease. Additionally, the Epstein-Barr virus (EBV) is etiologically linked to Burkitt's lymphoma and nasopharyngeal carcinoma. However, perhaps one of the most interesting chapters in herpes virus diseases is the relationship between HSV-2 and cervical carcinoma.

More than 15 years ago when Jeang was a graduate student studying herpes viruses it was an accepted dogma that HSV-2 was the persuasive etiological agent for cervical carcinoma. A plethora of seroepidemiological[20] studies supported by gene transfer experiments[31] led to the belief that HSV-2 viral DNA had transforming potential which explained the clinical development of cervical carcinoma in affected women. The strong belief that HSV-2 accounted for cervical cancer despite difficulties in demonstrating viral DNA in cervical cancers led to the proposition of a "hit-and-run" hypothesis.[17] This hypothesis, that a virus would transform cells and yet leave no residual footprint of its genetic presence, was designed to explain why "HSV-2-transformed" cells do not contain viral DNA and why the presence of viral DNA might not be necessary for the maintenance of a transformed phenotype. Thus, at the time, a complex virus-cell interaction seemed to have a relatively unambiguous answer.

Although the HSV-2 data explained many of the biological variables known to cervical carcinoma at a particular point in time, they did not (and could not) account for unknown variables. For instance, in the early 1980s, zur Hausen and colleagues discovered that although HSV-2 DNA could not be correlatively linked with cervical carcinomas, human papillomavirus (HPV) DNA was present in a very high percentage of cervical cancers and genital papillomas.[13,19] With this new finding, a change in the viral etiology for cervical carcinoma was suggested. Current understanding of viral etiologies now support that HPV plays a more significant role than HSV-2 in disease development.[18] Retrospectively, cervical cancer is highlighted by its strong epidemiological association with smoking, age, and multiple sexual partners. Whether there are other unknowns that will reshape our interpretations await further investigation.

HERPES VIRUS AND KAPOSI'S SARCOMA

Kaposi's sarcoma (KS) illustrates another case in which careful consideration of viral etiologies is important to understanding complex pathologies. Kaposi's sarcoma was first described in 1872 by a Hungarian physician (Mauritz Kaposi Kohn) as a rare, slowly progressive neoplasm in elderly men. In the 1960s, it became apparent that KS is endemic in parts of Africa and that the African variant shows an aggressively pathological phenotype in young people. In the 1980s, KS emerged as the most common HIV-associated malignancy with 30–50% penetrance in AIDS patients (reviewed in Miles[27]). However, despite its long history, until the mid-1990s, the etiological basis for KS remained unclear.

The association of KS with HIV prompted many investigators to study the molecular basis for this disease. Indeed for AIDS-associated KS, a series of reports indicated roles for the HIV-1 Tat protein and a variety of angiogenic growth factors[2,4] in disease development and maintenance. The data supporting these etiologies were substantial; yet the explanations were not completely satisfying since they did not provide a unified explanation for non-HIV-1-associated disease. In late 1994, work based on representational difference analysis from Chang et al.[9] offered groundbreaking evidence that KS is etiologically linked to a new virus—Kaposi's sarcoma herpes virus (KSHV), more recently reclassified as human herpes virus 8 (HHV8). HHV8, since that time, has been isolated and successfully passaged in tissue culture.[7,30] A wealth of new molecular epidemiological studies now confirms a linkage between HHV-8 and all types of HIV- and non-HIV-associated Kaposi's sarcoma,[23,29] although definitive cause and effect remain to be fully established.

The story of HHV-8 and KS represents yet another example in which a clear explanation for complex virus-cell interactions required clarification of variables that were not known at a particular juncture in time. Thus, this and other historical anecdotes reinforce the concept that provisional solutions that do not explain fully all variables in viral pathogenesis should be regarded and clearly labeled as answers-in-progress. Retrospective analyses are facile, and while we suggest constructive caution in data interpretation we do not intend to convey a destructive criticism of earlier analyses.

MANY COMPLEX QUESTIONS IN THE STUDY OF HIV

Equally difficult issues confront biological studies of HIV. HIV-1 is difficult to study because there is not a good and well-accepted animal model for this virus. Hence, many different "next-best" approaches (each may not be equivalent nor comparable to the other) have been used to derive answers. For instance, results from cell lines might not be comparable with data obtained from transgenic animal models; and infections done in humanized SCID-mice might not be equivalent to those in monkeys. Despite the obviousness of this statement, the HIV literature does contain interesting and provocative findings from many disparate systems that require much reconciliation. A part of the challenge in understanding HIV complexities is to decipher those inherent to virus-host interactions and those that arise from system-to-system differences. Below, we discuss some of these questions.

HIV infection and disease progression are complex processes which involve viral and host cell determinants. Tissue culture models allow us to study some aspects of the interaction between the virus and the cell; however, this approach as a disease system is obviously limited by the absent role of the immune system. Furthermore, the traditional tissue culture experiment has two significant limitations. First, more often than not, the cellular substrates for the virus are transformed T-cell lines. For instance, two of the more popular laboratory cell lines used in HIV-1 infections are MT-2 and MT-4 cells. Both are HTLV-I transformed cells containing HTLV-I genome and gene products. In many settings, the role of HTLV-I does not become an issue; however, in some cases this might unknowingly contribute a significant effect. Second, there is a practical question as to the strain of HIV-1 used in tissue culture experiments. Viruses that are well adapted to culture environments give robust infections and are easy to propagate. Although these viruses can provide quick readouts, in some cases, the answers are misleading. Hence, in the simplified tissue culture system, some viral genes (e.g., vpr, nef, vpu) have been shown to be nonessential for replication. However, more refined essays from other models and the observation that viruses in nature maintain open these reading frames suggest that the *ex vivo* culture results do not accurately reflect *in vivo* requirements.

One specific example illustrating interpretative complexities in understanding biological functions is the vif gene product. Vif (virion infectivity factor) is a late viral protein whose role in the virus life cycle is incompletely understood. Besides a role in particle assembly, Vif has been postulated to be involved in other steps such as reverse transcription, incorporation of the virus envelope, and efficient internalization of Env glycoprotein.[6,32] As if this myriad of functions is not complicated enough, another curious feature to Vif is that depending on the cell line used to produce the viral particles and to propagate subsequent reinfections, a $vif(-)$ virus can look like a $vif(+)$ virus in replication phenotype. Thus, some cells are "Vif-permissive" and others are "Vif-nonpermissive." In nonpermissive cells (e.g., H9, CEM, PBMC, macrophages) Vif is essential for virus replication and $vif(-)$ virus produced from these cells cannot reinfect either permissive or nonpermissive cells.[5,16,32] On the other hand, when $vif(-)$ virus is produced from permissive cells (e.g., HeLa, SupT1, Jurkat, C8166) and used to reinfect another permissive cell line, the replication phenotype resembles a $vif(+)$ virus. If the same thusly produced $vif(-)$ virus is used to reinfect a nonpermissive cell line, then only a single cycle of reinfection is permitted.[16,32] Thus, depending on the order-of-addition or the particular choice of cell lines, apparently disparate "true" results can emerge. Many explanations are possible; the most likely is that of a contribution from a yet to be identified host-cell factor.

Another gene product that has engendered complex interpretations is Nef (negative factor). Early on, it was noted that $nef(-)$ HIV-1s replicated very well in tissue culture, perhaps even a little bit better than $nef(+)$ viruses. This led to the proposition that the function of Nef was to repress transcription from the HIV-1 LTR.[1,26] According to this perspective, Nef affects virus replication negatively, and removal of this negative factor would positively augment virus growth. Curiously, when this hypothesis was tested in a nonhuman primate model, the opposite was found. In 1991, Kestler and colleagues infected rhesus monkeys with an SIVmac virus containing a premature stop codon in $nef.$[24] When monkeys were thusly inoculated, contrary to the expectations of Nef being a negative factor, quick revertant viruses that "cured" the stop codon were recovered. Thus, *in vivo*, Nef appears to be a positively and powerfully selected function. Retrospectively, reexaminations of *ex vivo* tissue culture infections show that in some cells Nef is clearly a positive viral infectivity factor.[11,28]

For some important functions, eukaryotic cells have developed a system of redundancies to guard against deleterious mutations. Viruses, too, have incorporated redundant functions, and this property adds complexity in interpreting results. An example of redundancy is found in the complementing roles of Vpr and MA (matrix protein) in the nuclear translocation of HIV-1 preintegration complexes. Lentiviruses, differing from oncoviruses, have a mechanism used to translocate genetic material from the cytoplasm into the nucleus of nondividing cells. In trying to understand the molecular determinants for this function, Bukrinsky and colleagues[8] used an infectious molecular clone derived from IIIB to demonstrate the role of MA in the nuclear transport of the preintegration complex. They showed that the basic domain of MA encoded a translocation function in nondividing cells. Interestingly, the IIIB molecular genome used in that study was fortuitously mutated in Vpr. This became an interpretative issue since in the setting of $vpr(+)$ genomes others[15] could not demonstrate that a mutation in the basic domain of MA had a deleterious phenotype for nuclear translocation of preintegration complexes. The conflicting findings were partially reconciled by the finding that Vpr has a redundant function in the entry of preintegration complexes into the nucleus. Hence, either MA or Vpr can dictate the movement of preintegrated HIV-1 DNA from the cytoplasm into the nucleus, and the effects of the two proteins are synergistic when they act together.[22]

The recent discovery of the HIV-1 coreceptor offers another example that complex cellular determinants contribute to difficult interpretations of virus biology. Work from many laboratories has shown that a seven-transmembrane-domain chemokine receptor CK-R5 (C-C chemokine receptor-5) which mediates the in-flammatory response of T- and phagocytic-cells to MIP-1α, MIP1-β, and RANTES serves as a coreceptor for macrophage-tropic HIV-1s.[3,10] With the elucidation of this function, investigators have demonstrated the existence of null alleles for CK-R5 which render cells resistant to infection by macrophage-tropic strains. Thus, Liu et al.[25] and Samson et al.[33] showed that null alleles for CK-R5 exist in the Caucasian population but not in Blacks from Western and Central Africa nor in the Japanese population. Moreover, epidemiological surveys are consistent with homozygous mu-tations in CK-R5 protecting individuals from HIV-1 infection. The fact that pro-tective CK-R5 mutations are racially restricted suggests that there are likely to be other host-genetic factors determining HIV-1 disease progression. Indeed, studies of West African prostitutes would support this idea. Thus emerges another level of complexity. Beyond cell-to-cell and virus-to-virus differences, people-to-people dif-ferences must also be considered in interpreting the biological and epidemiological manifestations of AIDS.

Issues concerning population-to-population differences are frequently approach-ed using epidemiological data. Epidemiology is a powerful and complex statistical tool for understanding medical etiologies. For example, Farr, a physician in the Victorian period, developed a mathematical model demonstrating the correlation between a factor, i.e., proximity to the Thames River in London, and the occur-rence of cholera. At that time, *Vibrio cholera*, the etiological agent, had not been discovered. Farr believed that cholera was a consequence of miasma, whereby peo-ple became sick when they inhaled this vapor. He searched for parameters which influenced the occurrence of disease and finally established a firm correlation be-tween the elevation of the Thames River where people lived and the mortality rate of cholera. Thus, he concluded that people who lived at higher elevations from the level of the river water had lower rates of cholera. For Faar, the level of the water was the raison d'etre for disease.[14] This was, of course, misleading as the later dis-covery of *Vibrio cholera* demonstrated the etiological organism directly responsible for this waterborne disease.

Epidemiological studies can contribute to progress in understanding diseases, but if injudiciously used and interpreted they can produce false leads. One extreme and facetious example is the study in 1988 by Peto and colleagues on the effect of aspirin in protecting against a second heart attack. When the study was first submitted to a highly reputed journal, Peto was asked to clarify the effects of aspirin on different subgroups of patients in order to identify variations in patients' response to aspirin.[12] Peto argued that subdividing into too many groups can lead to random correlations between irrelevant factors; and he illustrated this point by showing that depending on the patient's zodiac sign, aspirin is or is not beneficial in preventing a second heart attack (as recounted in a *Washington Post* column by Rick Weiss).

In viral diseases, epidemiology has an unquestionably important role. For AIDS, many examples of difficult-to-interpret epidemiogical findings exist. One challenging objective is to understand clearly how some groups of individuals might (might not) be protected from disease. Recently, Travers and colleagues[34] showed that prior infection with HIV-2 resulted in protection against subsequent HIV-1 infection. Interestingly, the same data was reinterpreted by Greenberg and colleagues[21] who reached a very different conclusion. There are mathematical manipulations and technical details that lead to two interpretations, but the fact that totally opposite conclusions can be derived from the same data set reinforces the need for caution in assigning certainty to answers.

CONCLUDING REMARKS

Understanding virus-induced diseases, either HIV or non-HIV, is compounded by the complexities inherent to the virus, the host reaction, epidemiological considerations, study-system variables, and, indeed, virus superinfections of the primary viral disorder (e.g., CMV and HHV-8 in HIV). It is perhaps not unexpected that complex biological questions should have equally complex answers. In some cases, virological processes are poorly amenable to reductive analyses. Issues posed by an integral system may not be approachable by simplified restructuring of the parts within the whole. Thus, reductive paradigms can lead to stepwise answers; however, the sum of the steps may not always lead to the answer for the whole. Inability to separate a complex process into simpler parts leads to situations where multiple unknowns need to be solved simultaneously in order to reach a clear answer. While it is human nature to desire clear-cut solutions, in many instances, one should recognize that significant uncertainties are inherent to conclusions derived from the many limitations built into different experimental approaches. The anecdotal examples presented above illustrate some of the challenges in interpreting the complex biology of HIV and other viruses.

ACKNOWLEDGMENTS

We thank E. Rich, M. Benkirane, R. Chun, V. Giordano, D. Jin, and H. Xiao for critical readings of this chapter. Work in our laboratory is supported in part by the AIDS Targeted Antiviral Program from the Office of the Director, NIH.

REFERENCES

1. Ahmad, N., and S. Venkatesan. "Nef Protein of HIV-1 is a Transcriptional Repressor of HIV-1 LTR." *Science* **241** (1988): 1481–1485.
2. Albini, A., G. Fontanini, L. Masiello, C. Tacchetti, D. Bigini, P. Luzzi, D. M. Noonan, and W. G. Stetler-Stevenson. "Angiogenic Potential *In Vivo* by Kaposi's Sarcoma Cell-Free Supernatants and HIV-1 Tat Product: Inhibition of KS-like Lesions by Tissue Inhibitor of Metalloproteinase-2." *Aids* **8** (1994): 1237–1244.
3. Alkhatib, G., C. Combadiere, C. C. Broder, Y. Feng, P. E. Kennedy, P. M. Murphy, and E. A. Berger. "CC CKR5: A RANTES, MIP-1α, MIP-1β Receptor as a Fusion Cofactor for Macrophage-Tropic HIV-1." *Science* **272** (1996): 1955–1958.
4. Barillari, G., R. Gendelman, R. C. Gallo, and B. Ensoli. "The Tat Protein of Human Immunodeficiency Virus Type 1, a Growth Factor for AIDS Kaposi Sarcoma and Cytokine-Activated Vascular Cells, Induces Adhesion of the Same Cell Types by Using Integrin Receptors Recognizing the RGD Amino Acid Sequence." *Proc. Natl. Acad. Sci., USA* **90** (1993): 7941–7945.
5. Blanc, D., C. Patience, T. F. Schulz, R. Weiss, and B. Spire. "Transcomplementation of VIF− HIV-1 Mutants in CEM Cells Suggests that VIF Affects Late Steps of the Viral Life Cycle." *Virology* **193** (1993): 186–192.
6. Borman, A. M., C. Quillent, P. Charneau, C. Dauguet, and F. Clavel. "Human Immunodeficiency Virus Type 1 Vif− Mutant Particles from Restrictive Cells: Role of Vif in Correct Particle Assembly and Infectivity." *J. Virol.* **69** (1995): 2058–2067.
7. Boshoff, C., D. Whitby, T. Hatziioannou, C. Fisher, J. van der Walt, A. Hatzakis, R. Weiss, and T. Schulz. "Kaposi's-Sarcoma-Associated Herpesvirus in HIV-Negative Kaposi's sarcoma." *Lancet* **345** (1995): 1043–1044.
8. Bukrinsky, M. I., S. Haggerty, M. P. Dempsey, N. Sharova, A. Adzhubel, L. Spitz, P. Lewis, D. Goldfarb, M. Emerman, and M. Stevenson. "A Nuclear Localization Signal Within HIV-1 Matrix Protein that Governs Infection of Nondividing Cells." *Nature* **365** (1993): 666–669.
9. Chang, Y., E. Cesarman, M. S. Pessin, F. Lee, J. Culpepper, D. M. Knowles, and P. S. Moore. "Identification of Herpesvirus-like DNA Sequences in AIDS-Associated Kaposi's Sarcoma." *Science* **266** (1994): 1865–1869.
10. Choe, H., M. Farzan, Y. Sun, N. Sullivan, B. Rollins, P. D. Ponath, L. Wu, C. R. Mackay, G. LaRosa, W. Newman, N. Gerard, C. Gerard, and J. Sodroski. "The Beta-Chemokine Receptors CCR3 and CCR5 Facilitate Infection by Primary HIV-1 Isolates." *Cell* **85** (1996): 1135–1148.
11. Chowers, M. Y., C. A. Spina, T. J. Kwoh, N. J. Fitch, D. D. Richman, and J. C. Guatelli. "Optimal Infectivity *In Vitro* of Human Immunodeficiency Virus Type 1 Requires an Intact Nef Gene." *J. Virol.* **68** (1994): 2906–2914.

12. Collins, R., R. Peto, S. MacMahon, P. Hebert, N. H. Fiebach, K. A. Eberlier, J. Godwin, N. Qisibash, J. O. Taylor, and C. H. Hennekens. "Blood Pressure, Stroke, and Coronory Heart Disease. Part 2. Short-Term Recutions in Blood Pressure: Overview of Randomised Drug Trials in Their Epidemiological Context." *Lancet* **335** (1990): 1534–1535.

13. de Villiers, E. M., D. Wagner, A. Schneider, H. Wesch, H. Miklaw, J. Wahrendorf, U. Papendick, and H. zur Hausen. "Human Papillomavirus Infections in Women With and Without Abnormal Cervical Cytology." *Lancet* **2** (1987): 703–706.

14. Eyler, J. M. *Victorian Medicine, the Ideas and Methods of William Farr.* Baltimore: Johns Hopkins University Press, 1979.

15. Freed, E. O., and M. A. Martin. "HIV-1 Infection of Nondividing Cells." *Nature* **369** (1994): 107–108.

16. Gabuzda, D. H., K. Lawrence, E. Langhoff, E. Terwilliger, T. Dorfman, W. A. Haseltine, and J. Sodroski. "Role of Vif in Replication of Human Immunodeficiency Virus Type 1 in CD4+ T Lymphocytes." *J. Virol.* **66** (1992): 6489–6495.

17. Galloway, D. A., and J. K. McDougall. "The Oncogenic Potential of Herpes Simplex Viruses: Evidence for a 'Hit-and-Run' Mechanism." *Nature* **302** (1983): 21–24.

18. Galloway, D. A., and J. K. McDougall. "Human Papillomaviruses and Carcinomas." *Adv. Virus Res.* **37** (1989): 125–171.

19. Gissmann, L., L. Wolnik, H. Ikenberg, U. Koldovsky, H. G. Schnurch, and H. zur Hausen. "Human Papillomavirus Types 6 and 11 DNA Sequences in Genital and Laryngeal Papillomas and in Some Cervical Cancers." *Proc. Natl. Acad. Sci., USA* **80** (1983): 560–563.

20. Graham, S., W. Rawls, M. Swanson, and J. McCurtis. "Sex Partners and Herpes Simplex Virus Type 2 in the Epidemiology of Cancer of the Cervix." *Am. J. Epidemiol.* **115** (1982): 729–735.

21. Greenberg, A. E., S. Z. Wiktor, K. M. DeCock, P. Smith, H. W. Jaffe, and T. Dondero Jr. "HIV-2 and Natural Protection Against HIV-1 Infection." *Science* **272** (1996): 1959–1960.

22. Heinzinger, N. K., M. I. Bukinsky, S. A. Haggerty, A. M. Ragland, V. Kewalramani, M. A. Lee, H. E. Gendelman, L. Ratner, M. Stevenson, and M. Emerman. "The Vpr Protein of Human Immunodeficiency Virus Type 1 Influences Nuclear Localization of Viral Nucleic Acids in Nondividing Host Cells." *Proc. Natl. Acad. Sci., USA* **91** (1994): 7311–7315.

23. Huang, Y. Q., J. J. Li, M. H. Kaplan, B. Poiesz, E. Katabira, W. C. Zhang, D. Feiner, and A. E. Friedman-Kien. "Human Herpesvirus-Like Nucleic Acid in Various Forms of Kaposi's Sarcoma." *Lancet* **345** (1995): 759–761.

24. Kestler, H., D. J. Ringler, K. Mori, D. L. Panicali, P. K. Sehgal, M. D. Daniel, and R. C. Desrosiers. "Importance of the Nef Gene for Maintenance of High Virus Loads and for Development of AIDS." *Cell* **65** (1991): 651–662.

25. Liu, R., W. A. Paxton, S. Choe, D. Ceradini, S. R. Martin, R. Horuk, M. E. MacDonald, H. Stuhlmann, R. A. Koup, and N. R. Landau. "Homozygous Defect in HIV-1 Coreceptor Accounts for Resistance of Some Multiply Exposed Individuals to HIV-1 Infection." *Cell* **86** (1996): 367–377.

26. Luciw, P. A., C. Cheng-Mayer, and J. A. Levy. "Mutational Analysis of the Human Immunodeficiency Virus: The Orf-B Region Down-Regulates Virus Replication." *Proc. Natl. Acad. Sci., USA* **84** (1987): 1434–1438.

27. Miles, S. A. "Pathogenesis of HIV-Related Kaposi's Sarcoma." *Curr. Opin. Oncol.* **6** (1994): 497–502.

28. Miller, M. D., M. T. Warmerdam, I. Gaston, W. C. Greene, and M. B. Feinberg. "The Human Immunodeficiency Virus-1 Nef Gene Product: A Positive Factor for Viral Infection and Replication in Primary Lymphocytes and Macrophages." *J. Exp. Med.* **179** (1994): 101–113.

29. Moore, P. S., and Y. Chang. "Detection of Herpesvirus-Like DNA Sequences in Kaposi's Sarcoma in Patients With and Without HIV Infection." *N. Engl. J. Med.* **332** (1995): 1181–1185.

30. Renne, R., W. Zhong, B. Herndier, M. McGrath, N. Abbey, D. Kedes, and D. Ganem. "Lytic Growth of Kaposi's Sarcoma-Associated Herpesvirus (Human Herpesvirus 8) in Culture." *Nat. Med.* **2** (1996): 342–346.

31. Reyes, G. R., R. LaFemina, S. D. Hayward, and G. S. Hayward. "Morphological Transformation by DNA Fragments of Human Herpesviruses: Evidence for Two Distinct Transforming Regions in Herpes Simplex Virus Types 1 and 2 and Lack of Correlation with Biochemical Transfer of the Thymidine Kinase Gene." *Cold Spring Harbor Symp. Quant. Biol.* **1** (1980): 629–641.

32. Sakai, H., R. Shibata, J. Sakuragi, S. Sakuragi, M. Kawamura, and A. Adachi. "Cell-Dependent Requirement of Human Immunodeficiency Virus Type 1 Vif Protein for Maturation of Virus Particles." *J. Virol.* **67** (1993): 1663–1666.

33. Samson, M., F. Libert, B. J. Doranz, J. Rucker, C. Liesnard, C. M. Farber, S. Saragosti, C. Lapoumeroulie, J. Cognaux, C. Forceille, G. Muyldermans, C. Verhofstede, G. Burtonboy, M. Georges, T. Imai, S. Rana, Y. Yi, R. J. Smyth, R. G. Collman, R. W. Doms, G. Vassart, and M. Parmentier. "Resistance to HIV-1 Infection in Caucasian Individuals Bearing Mutant Alleles of the CCR-5 Chemokine Receptor Gene." *Nature* **382** (1996): 722–725.

34. Travers, K., S. Mboup, R. Marlink, A. Gueye-Nidaye, T. Siby, I. Thior, I. Traore, A. Dieng-Sarr, J. L. Sankale, C. Mullins, V. I. Ndoye, C. C. Hsieh, M. Essex, and P. Kanki. "Natural Protection Against HIV-1 Infection Provided by HIV-2." *Science* **268** (1995): 1612–1615.

Index